Cambrian acritarchs from Upper Silesia, Poland – biochronology and tectonic implications

MAŁGORZATA MOCZYDŁOWSKA

Moczydłowska, M. 1998 10 31: Cambrian acritarchs from Upper Silesia, Poland – biochronology and tectonic implications. *Fossils and Strata*, No. 46, pp. 1–121. Oslo. ISSN 0300-9491. ISBN 82-00-37692-3.

Subsurface sedimentary successions in the Sosnowiec IG-1, Goczałkowice IG-1 and Potrójna IG-1 boreholes of Upper Silesia were studied for organic-walled microfossils in order to establish the biochronology of these otherwise sparsely fossiliferous strata. Taxonomically diverse associations of acritarchs allow recognition of the Lower, Middle and Upper Cambrian series and more detailed zones within the Lower and Middle Cambrian. In relation to the Baltic craton, the area is a suspect terrane located within the Trans-European Suture Zone. The taphonomy of microfossils and the thermal maturation of organic matter, in the context of sedimentary structures and facies associations, indicate that the microfossils occur *in situ* in the depositional settings. The distribution of morphotypes and sizes of microfossils in various depositional environments suggest that there is no selective distribution of microplankton and thus no distinctive plankton communities occupying nearshore and offshore shallow-shelf environments. The biodiversity of Cambrian phytoplankton is reviewed in relation to global and regional geoevents. Several major radiation and extinction events are recognized. The record from Upper Silesia is compared to the pattern of Cambrian secular biodiversity changes, revealing some regional variations at the Lower–Middle Cambrian boundary. The duration of the lowest Cambrian acritarch zones, the *Asteridium tornatum – Comasphaeridium velvetum* and *Skiagia ornata – Fimbriaglomerella membranacea* zones and their time-equivalent faunal zones of *Platysolenites antiquissimus* and *Schmidtiellus mickwitzi*, is estimated as 2–5 Ma each. The Upper Silesia terrane represents a distal portion of East Avalonia at the margin of Gondwana. The homogenous distribution of phytoplankton along the shelves of Baltica, Gondwana and Laurentia facing the Iapetus and Avalonian seaways indicates that these basins were still relatively narrow in Early Cambrian times, i.e. at the initial opening stages. The Upper Silesia terrane has been accreted to Baltica along the Kraków–Myszków Fault Zone, considered to be an extension of the Tornquist Suture and possibly coinciding with the Trans-European Fault. Seventy-five form-taxa are described, with synonymies and a comprehensive compilation of their stratigraphic ranges and geographic distribution. The taxonomic status of a number of taxa is revised, rejecting so-called 'polymorphic genera' and arbitrarily chosen dimensional limits in the diagnoses. The new genus *Duplisphaera* is recognized, as well as eighteen new species: *Adara undulata, Asteridium pilare, A. solidum, Celtiberium? papillatum, Comasphaeridium silesiense, Cymatiosphaera pusilla, Heliosphaeridium bellulum, H. exile, H. nodosum, H. serridentatum, Multiplicisphaeridium ramosum, M. sosnowiecense, M. varietatis, Solisphaeridium bimodulentum, S. cylindratum, S. elegans, Stelliferidium robustum,* and *Vogtlandia simplex*. Six additional taxa are left under open nomenclature. The diagnoses of two genera, *Revinotesta* and *Solisphaeridium*, and three species, *Solisphaeridium baltoscandium, S. flexipilosum,* and *S. multiflexipilosum*, are emended. Nine new combinations of species are proposed: *Duplisphaera luminosa, Heliosphaeridium lanceolatum, H. oligum, Lophosphaeridium latviense, Multiplicisphaeridum parvum, Polygonium varium, Revinotesta izhorica, R. saccata,* and *Solisphaeridium implicatum*. The general taxonomic concepts of acanthomorphic and polygonomorphic acritarch genera are discussed in the view of recent evaluations. □*Cambrian, acritarchs, organic-walled microfossils, taphonomy, taxonomy, biochronology, biodiversity, palaeogeography, tectonic history, Poland, Upper Silesia terrane, Cadomian basement, East Avalonia, Tornquist Suture.*

Małgorzata Moczydłowska [malgo.vidal@pal.uu.se], Institute of Earth Sciences, Micropalaeontology, Uppsala University, Norbyvägen 22, S-752 36 Uppsala, Sweden; 1st January, 1996; revised 28th January, 1997.

Contents

Introduction

The continental crust of Phanerozoic Europe was formed around an Archean–Proterozoic Baltica palaeocontinent by accretion of fold belts that developed during three orogenies, i.e. Caledonian, Variscan and Alpine, and by suturing of East Avalonian (Gondwana derived) and various suspect terranes. One of these terranes is the Upper Silesia area in southern Poland, now located in the vicinity of the Caledonian Deformation Front (=Teisseyre–Tornquist Zone) and the Trans-European Fault. The Upper Silesia terrane is fault-bounded and flanked by fold belts, yet it has remained unfolded and unmetamorphosed since Early Cambrian or Vendian times.

The subject of this study is the lowermost segment of a sedimentary succession underlying the sub-Devonian unconformity and unconformably overlying metamorphosed and deformed basement complexes. Within a 50 m thick interval, the succession contains rare olenellid trilobites of the Acado-Baltic faunal province indicating an Early Cambrian age. The remaining approximately 350 m of strata are devoid of any age-diagnostic fossils and were examined for acritarchs. Acritarchs are organic-walled microfossils interpreted as planktic photoautotrophic protoctists that occur in marine sedimentary rocks spanning at least ca. 1.7 Ga of Earth history (i.e. Proterozoic to Holocene). They have been used successfully for biostratigraphic purposes, particularly in Lower Palaeozoic and Neoproterozoic strata. In addition to their significance for revealing the early evolution of biosphere and marine ecosystems, acritarchs can contribute substantially to the recognition of the palaeobiogeographic affinities of presently dismembered terranes (fragments of palaeocontinents) and their tectonic history.

The present study revealed that taxonomically diverse and well-preserved acritarchs occur throughout the Upper Silesian succession, indicating the presence of the Lower, Middle and Upper Cambrian series. Age-diagnostic acritarch associations have enabled stratigraphic relationships within the investigated Cambrian strata to be established, demonstrating a normal geological succession and low to moderate thermal alteration. The biochronology of the succession constrains the relative age of the deformation and metamorphism of the underlying metasedimentary basement sequence, which can be referred to the Cadomian orogeny. It also brackets the age of molasse deposits within the succession, here regarded as post-Cadomian.

The palaeogeographic location of the Upper Silesia terrane in Cambrian times as a distal segment of East Avalonia is inferred from the taxonomic similarity of the present acritarch associations to those reported from Baltica, Avalonia and Armorica. This in turn sets some constraints upon the tectonic development of the pre-Old Red Sandstone sequence in the Upper Silesia terrane. As a result of plate motions, the Upper Silesia terrane, once part of a remote margin of Gondwana, is now located in the centre of the European continent. From this point of view it is a unique succession of fossils and strata.

Material and methods

Sedimentary successions penetrated by the deep research boreholes Sosnowiec IG-1, Goczałkowice IG-1 and Potrójna IG-1 in Upper Silesia were examined for organic-walled microfossils (Figs. 1, 2). Samples from the

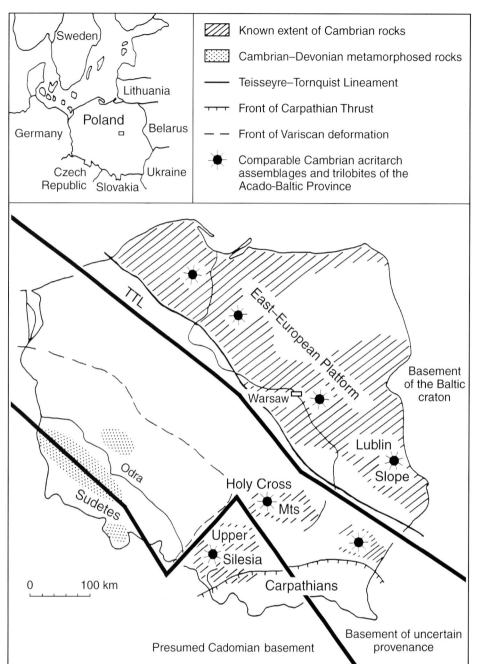

Fig. 1. Geologic sketch-map of Poland showing the investigated area in Upper Silesia and extension of Cambrian rocks in terranes with Precambrian basement of various provenances. Distribution of acritarch assemblages is according to Moczydłowska (1991) and present data. Modified after Moczydłowska (1995a, b).

cored intervals were collected, and their depths are given in metres below ground level according to the core logs. Core recovery of Precambrian–Cambrian strata was approximately 45% in the Sosnowiec and Goczałkowice boreholes, and 29% in the Potrójna borehole. However, the portions of the cores suitable for sampling, with respect to state of preservation and lithology, were very limited. All samples are from organic-rich, fine-grained siliciclastic rocks (dark grey in colour), comprising mudstones and siltstones, or mudstones alternating with fine-grained sandstones. In the Sosnowiec succession, samples

from twelve different stratigraphic intervals were examined and eleven of them were fossiliferous. They contain abundant and generally well-preserved acritarchs and are rich in particulate organic matter. Four samples out of fifteen from the Goczałkowice borehole were found to be barren, and one of the fossiliferous samples yielded only undiagnostic sphaeromorphs. The frequency and taxonomic diversity of microfossils is much lower than in the Sosnowiec succession. Amorphous fragments of kerogen observed in microscopic slides are very numerous. In the Potrójna succession samples from three stratigraphic lev-

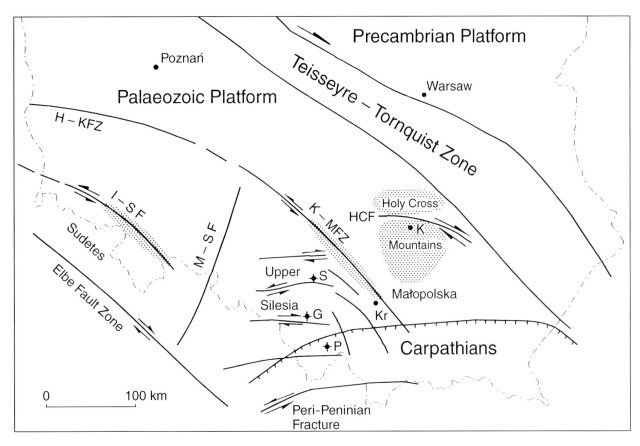

Fig. 2. Tectonic sketch-map of the Upper Silesia terrane with location of the investigated boreholes and distribution of major faults (solid lines) and areas of Caledonian deformation (shaded) in southern Poland. Line with dentation indicates front of the Carpathian Thrust, arrows indicate the direction of displacement along faults. H–KFZ = Hamburg–Kraków Fault Zone, K–MFZ = Kraków–Myszków Fault Zone, HCF = Holy Cross Fault, I-SF = Intra-Sudetic Fault, M–SF = Moravian–Silesian Fracture, K = Kielce, Kr = Kraków, S = Sosnowiec IG-1 borehole, G = Goczałkowice IG-1 borehole, P = Potrójna IG-1 borehole. The map shows that the Upper Silesia terrane appears to be a rigid crustal block between areas affected by Caledonian deformation. It has been further suggested that the Kraków–Myszków Fault Zone could be an offset extension of the Intra-Sudetic Fault, i.e. the Tornquist Suture as interpreted by Johnston *et al.* (1994). Compiled from Johnston *et al.* (1994), Żaba (1994), Kotas (1982a), Brochwicz-Lewiński *et al.* (1983b), and modified after Moczydłowska 1997).

els were studied, and scarce and poorly preserved acritarchs were recorded in two of them. Organic matter was also spare.

Conventional palynological preparation techniques, using 10% HCl, 45% HF and 62% HNO_3 for the digestion of the mineral components were used to separate the organic-walled microfossils (cf. Vidal 1988). The weight of the processed samples was 100 g, spent acid was decanted, and the residue was rinsed with deionized water. The organic residue was concentrated by centrifugation only at the final stage of rinsing with concentrated ethanol and acetone. Permanent strew mounts were produced using the synthetic resin Epotek 301 as a mounting medium. Two microscopic slides from each sample were examined under a transmitted light microscope (Leitz Wetzlar Dialux 20) provided with an interference contrast attachment.

Geological setting

General geology

The area of Upper Silesia in southern Poland is bordered by deep fractures and fault zones comprising the Kraków–Myszków Fault Zone (being part of the Hamburg–Kraków Fault Zone), the Peri-Peninian fracture zone, the Elbe Fault Zone, and the Moravian–Silesian fracture zone (Kotas 1982a; Brochwicz-Lewiński *et at.* 1983; Żaba 1994, 1995; Fig. 2 herein). The sedimentary successions in Upper Silesia dip gently and are unmetamorphosed. They are not affected by tectonic deformation other than block faulting, and only minor Hercynian folding affects Upper Palaeozoic sedimentary rocks in the northwestern margin of the area (Kotas 1982a). The sedimentary successions consist of Cambrian to Recent rocks with numerous stratigraphic hiati. They constitute a platform cover overlying plutonic and metamorphic basement rocks (Kotas 1982a; Ślączka 1985c; Brochwicz-Lewiński *et al.* 1986; Moczydłowska 1993a, 1995a, 1996a). This lack of deformation is in sharp tectonic contrast to the surrounding areas that constitute the fold belts and/or strongly deformed fault zones (Moczydłowska 1996a). The fold belts were formed during three orogenies and comprise the Holy Cross Mountains (Caledonian folding and/or thrusting), the Sudetes Mountains (Caledonian and Variscan), and the Carpathian Mountains (Alpine) (Kotas 1982a; Ślączka 1985c; Kowalczewski & Migaszewski 1993; van Bremmen *et al.* 1988; Oliver *et al.* 1993; Johnston *et al.* 1994; Dadlez *et al.* 1994; Fig. 2). The Kraków–Myszków zone, adjacent to the northeastern margin of Upper Silesia, has been interpreted either as an orogenic belt that was deformed during the Caledonian and Variscan orogenies (Znosko 1965; Bukowy 1982; Harańczyk 1982, 1994) and called the Krakowian Belt (Oliver *et al.* 1993; Johnston *et al.* 1994), or as a fault zone (Brochwicz-Lewiński *et al.* 1983). Brochwicz-Lewiński *et al.* (1983) interpreted this zone as a transpressive fault zone that was active during various periods of time from Early Ordovician to Late Carboniferous or ?Early Permian. Accordingly, the observed deformation of Palaeozoic sedimentary rocks in this zone could be the result of movements along the fault system.

The basement in Upper Silesia consists of plutonic and metamorphic complexes and a sequence of folded terrigenous deposits weakly metamorphosed in greenschist facies (Kotas 1982a). The latter metasediments, constituting pelitic and psammitic rocks, are overlain unconformably by unmetamorphosed and almost flat-lying Cambrian strata (Piotrowice 1 borehole, Brochwicz-Lewiński *et al.* 1986; Goczałkowice IG-1 and Potrójna IG-1 boreholes, see below). The crystalline basement rocks have been penetrated by several boreholes and coincide with magnetic and gravity anomalies (Heflik & Konior 1974; Ślączka 1976, 1985c; Heflik 1982; Brochwicz-Lewiński *et al.* 1986; Kowalczewski 1990). Gabbros (the Andrychów 3 borehole) and metamorphic granitoids and schists (the Bielsko 4, 5, Kęty 7, 8, 9, Andrychów 4 boreholes) occur in the southern part of Upper Silesia where they form the Bielsko-Andrychów horst. Gneisses, diabases and amphibolites extend southwards from the horst area to the Carpathians and eastwards to the Małopolska area (Łodygowice and Rzeszotary boreholes, respectively). However, the stratigraphic and tectonic relationships between the metasediment, igneous and strongly metamorphosed basement complexes are not clearly established. The contacts between these complexes have never been observed in boreholes, and there are no recent isotopic datings. Evidence to infer the stratigraphic succession of the discrete complexes is provided by the occurrence of pebbles of gabbro in metaconglomerates within a weakly metamorphosed metasediment sequence at Goczałkowice (Cebu-

lak *et al.* 1973a). It was inferred that metasediments unconformably overlie plutonic and strongly metamorphosed complexes (Kotas 1982a). Based on studies of petrography and grades of metamorphism, Heflik & Konior (1974) and Heflik (1982) suggested that gabbros and metasomatic granitoids intruded into paragneisses and schists. The original rocks metamorphosed into paragneisses and schists were probably sedimentary rocks (sandstones and shales), whereas previously metamorphosed sediments were metasomatically altered into granitoids (Heflik 1982). The depositional age of the sedimentary rocks, the age of their deformation and metamorphism, and the timing of intrusion of plutonic rocks are unknown, other than being Precambrian and preceding formation of the metasedimentary sequence.

The Upper Silesia area was considered to have formed a massif (the Upper Silesian Massif; Kotas 1982a; Bukowy 1982; Znosko 1984; Brochwicz-Lewiński *et al.* 1986), and metamorphism of the Precambrian basement has been attributed to the Cadomian event (Brochwicz-Lewiński *et al.* 1981, 1983, 1986). The Upper Silesia area has subsequently been regarded as a microcontinent (Bukowy 1984), suspect terrane (Brochwicz-Lewiński *et al.* 1986), or terrane (Pożaryski 1990, Nawrocki 1990; Oliver *et al.* 1993; Harańczyk 1994; Bukowy 1994; Dadlez 1995). As its tectonostratigraphic development is distinct from those of adjacent areas, the Upper Silesia area is regarded here as a terrane as understood by Oliver *et al.* (1993) and Stone & Kimbell (1995), being a part of East Avalonia (Moczydłowska 1995a, b, 1996a).

The sedimentary successions in the Upper Silesia area consist of several distinct depositional and stratigraphic units attributed to the Caledonian, Variscan and Alpine orogenic cycles. These units are defined by major regional unconformities associated with substantial stratigraphic hiati (Kotas 1982a; Ślączka 1985c). Additionally, the Carpathian nappes are thrust over autochthonous strata along the southern margin of the area (e.g., in the Potrójna borehole). The nappes consist of repeated successions of Jurassic, Cretaceous and Miocene sedimentary strata. The autochthonous sedimentary cover overlies the basement complexes with profound angular unconformity.

The Lower Palaeozoic unit is defined by the sub-Cambrian and sub-Devonian unconformities and comprises Cambrian transgressive siliciclastic sequences that are gently dipping and unmetamorphosed. There are neither Ordovician nor Silurian deposits in the area (Kotas 1982a; Bukowy 1982), but they occur in the adjacent Kraków–Myszków zone (Harańczyk 1982, 1994). The Cambrian successions are approximately 550 m in thickness and are only affected by faulting and gentle tilting.

The Upper Palaeozoic unit embraces Devonian and Carboniferous rocks (Kotas 1982a). The Lower Devonian deposits comprise the Old Red Sandstone facies, whereas

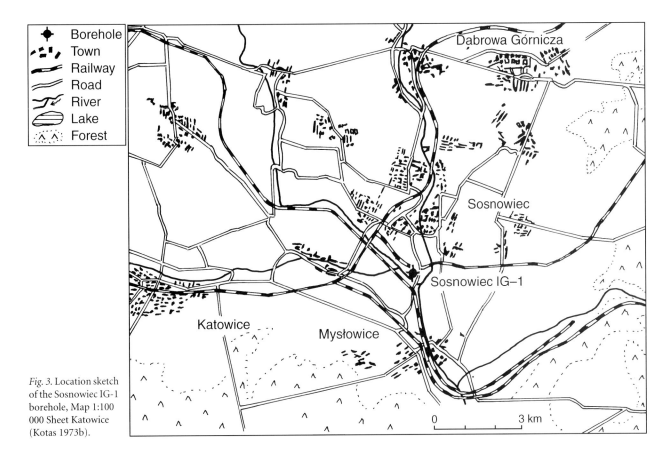

Fig. 3. Location sketch of the Sosnowiec IG-1 borehole, Map 1:100 000 Sheet Katowice (Kotas 1973b).

Legend:
- Borehole
- Town
- Railway
- Road
- River
- Lake
- Forest

Dabrowa Górnicza

Sosnowiec

Sosnowiec IG–1

Katowice   Mysłowice

0   3 km

the Middle and Upper Devonian and lower part of the Lower Carboniferous consist of carbonate platform facies. These rocks are overlain with sedimentary break, reflecting the Bretanian orogenic event, by a flysch-like association of sediments referred to the upper part of the Lower Carboniferous and the Upper Carboniferous. They are succeeded, following a transition, by molasse deposits of the upper part of the Upper Carboniferous (Kotas 1982a). All these sequences are around 7000 m thick and form the Upper Silesian Carboniferous Coal Basin. The Hercynian tectonic deformation affecting rocks in Upper Silesia was mostly block-faulting and gentle warping in the central areas, whereas true but minor folding occurred in the western and northwestern areas adjoining the Variscan orogenic belt (Kotas 1972, 1982a). An unconformity and stratigraphic hiatus separates Carboniferous rocks from the overlying Permian, Triassic and Jurassic sediments, which form a monoclinal cover. Also unconformably overlying are autochthonous Tertiary (Miocene and Oligocene) strata developed in molasse facies. The Western Carpathian nappes are thrust over these successions along the southern margin of the area .

The successions examined in the present study occur in different parts of the Upper Silesia area (Fig. 2) and represent various parts of the Cambrian, including the Lower Cambrian, Middle Cambrian and Upper Cambrian

Series. They are known only from boreholes. The Sosnowiec IG-1 borehole is situated at Sosnowiec (lat. 50°15'50"N and long. 10°08'23"E ; Map 1:100 000 Sheet Katowice; Kotas 1973b; Fig. 3). The Goczałkowice IG-1 borehole was drilled in Goczałkowice Zdrój (lat. 49°56'13"N and long. 18°57'45"E; Map 1:100 000 Sheet Bielsko–Biała; Kotas 1973a; Fig. 4). The Potrójna IG-1 borehole is located near to Jaszczurowa, about 20 km south of the Carpathian Thrust front (Map 1:100 000 Sheet Bielsko–Biała; Ślączka 1976, 1985c; Fig. 5).

The Sosnowiec succession

The Cambrian succession in the Sosnowiec IG-1 borehole underlies the basal Devonian unconformity at a depth of 3156.0 m and extends to a depth of 3442.6 m (Kotas 1973b, c). The Cambrian rocks are referred to the informal Sosnowiec formation (Kowalczewski 1990), a terrigenous unit of alternating mudstones and fine-grained sandstones with minor interbeds of quartzitic and polymodal sandstones, conglomeratic at their base (Fig. 6), about 195 m thick. The succession is intruded by a gabbro-diorite and diabase sill attaining a thickness of 92 m (Cebulak *et al.* 1973b). The sedimentary rocks are strongly lithified, and the sandstones and mudstones are

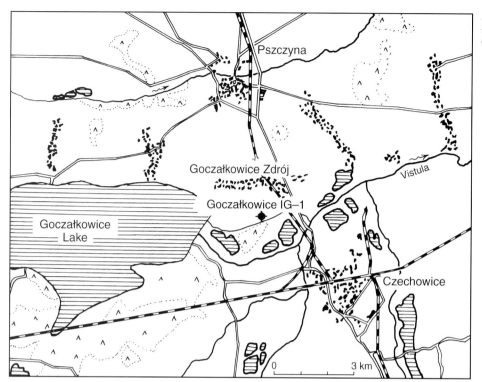

Fig. 4. Location sketch of the Goczałkowice IG-1 borehole, Map 1:100 000 Sheet Bielsko–Biała (Kotas 1973a). For legend see Fig. 3.

partly recrystallized, a feature that is observed not only in the vicinity of the above-mentioned igneous intrusion (Cebulak *et al.* 1973c, Kotas 1973b).

The basal portion of the Sosnowiec formation, between 3425.0 and 3442.6 m, was referred to the Radocha member by Kowalczewski (1990). It consists of fining-upwards conglomeratic and greywacke sandstones grading into the polymodal quartz sandstones. The sandstones are poorly sorted and massively bedded with sporadic draped clay-

stone laminae (Cebulak *et al.* 1973b). The Radocha member gradually passes into fine-grained quartzitic and muddy sandstones with mudstone intercalations (3392.0–3425.0 m) that are slightly bioturbated (Fig. 6). Inconspicuous parallel bedding and wave lamination occur within the mudstones, whereas shallow erosional scours occur occasionally in the otherwise predominantly massive sandstones. Additional minor sandstone beds, 10–15 m thick, occur higher in the succession underlying the intrusion and at the top of the formation. The bulk of the Sosnowiec formation is formed by a monotonous succession of mudstones and alternating mudstones and fine-grained sandstones that are well sorted, well bedded and in part thinly laminated. The beds are almost flat-lying, generally dipping at an angle of 2–7° (Kotas 1973b). The sedimentary structures observed at a few intervals are small-scale cross-bedding, graded bedding and flaser lamination. Simple trace fossils, consisting mostly of vertical and oblique burrows, occur throughout the strata from a level immediately above the Radocha member (Figs. 6 and 7). Fragments of shells and moulds of inarticulate brachiopods (lingulids and acrotretids) have been recorded in the lower part of the formation (Biernat & Baliński 1973; Kotas 1973b). Organic-walled microfossils and trace fossils are satisfactorily preserved, although the former are thermally altered. The Sosnowiec formation is considered to have accumulated in shallow marine environments comprising the tidal zone (the lower part of the formation) and the shelf below wave-base (Kotas 1973b).

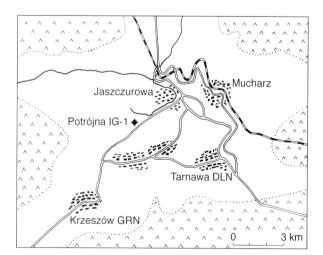

Fig. 5. Location sketch of the Potrójna IG-1 borehole, Map 1:100 000 Sheet Bielsko–Biała (Ślączka 1985c). For legend see Fig. 3.

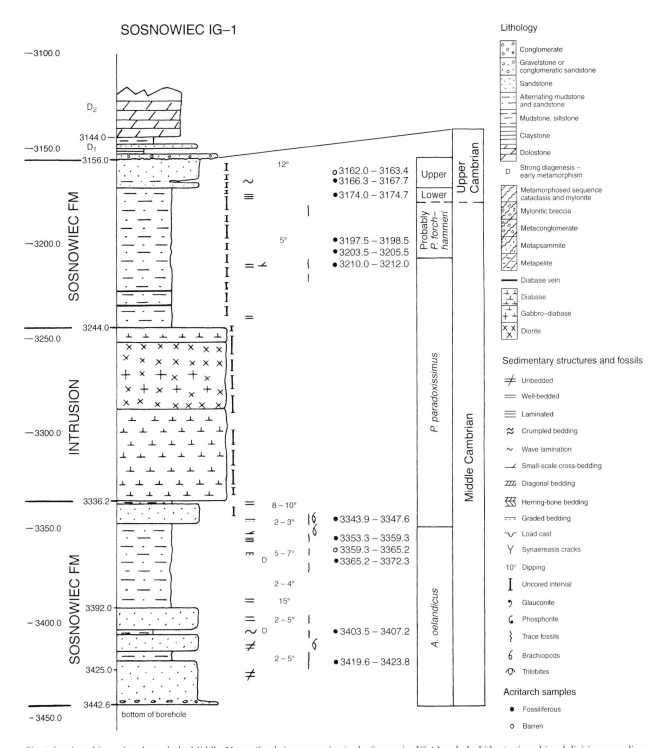

Fig. 6. Stratigraphic section through the Middle–Upper Cambrian succession in the Sosnowiec IG-1 borehole. Lithostratigraphic subdivision according to Kotas (1973b) and Kowalczewski (1990). Occurrence of brachiopods according to Biernat & Baliński (1973) and Kotas (1973b). Additional observations on sedimentary structures and occurrence of trace fossils according to this paper. D1 = Lower Devonian, D2 = Middle Devonian.

The Sosnowiec formation is unconformably overlain by Lower Devonian conglomeratic sandstones, mudstones and claystones succeeded by Middle Devonian dolostones. The Devonian strata, both detrital and carbonate, are unfossiliferous, and their relative age is inferred from lithostratigraphical correlations (Kotas 1973b). The sub-Devonian unconformity is erosional and apparently parallel (Kotas 1973c), or possibly a low angular unconformity. The overlying Devonian and Lower Carboniferous carbonate successions dip 2–10° and are

Fig. 7. Simple filled burrows of deposit feeders from the Sosnowiec IG-1 borehole. □A, B, D. Fragment of drillcore from interval 3359.3–3365.2 m, Sosnowiec formation, Middle Cambrian, *Acadoparadoxides oelandicus* Superzone. PMU-Pl.163. □A. Upper bedding surface showing portion of burrow (0.5 cm in diameter) parallel to bedding plane. □B, D. The same specimen, oblique (B) and side (D) views showing burrow turning from bedding surface and penetrating deeply into sediment. □C. Specimen on upper bedding surface from interval 3217.3–3218.3 m, Sosnowiec formation, Middle Cambrian, *Paradoxides paradoxissimus* Superzone. PMU-Pl.164. Diameter 0.8 cm, length 5.0 cm. Scale bar in B equals 1.3 cm for A, B, D; 1.0 cm for C.

generally concordant with Cambrian strata of the Sosnowiec formation (Kotas 1973b).

The magmatic intrusion in the sedimentary succession is a composite sill formed during two emplacement phases of gabbro-diorites (older phase in the central part of the sill) and diorites and diabases (younger phase in the outer zones) (Cebulak *et al.* 1973b). Thin diabase apophyses occur above the sill at 3223.6–3224.6 m and 3230.5–3231.5 m (Kotas 1973b). The contact zones between igneous rocks and sedimentary rocks of the Sosnowiec forma-

tion are thermally and hydrothermally altered, strongly recrystallized and silicified into hornfels (Kotas 1973b, Kowalczewski 1990). Xenolithic fragments of quartzitic sandstones occur in the vicinity of the apophysis at a depth of 3230.0–3231.5 m (Kotas 1973b). However, metamorphic aureoles are very local, being limited to a few metres above and below the sill. The age of emplacement of the sill and apophyses is unknown since there is no isotopic dating on the igneous rocks, either in this succession or in the area in general.

The entire sedimentary succession below the sub-Devonian unconformity in the Sosnowiec borehole was previously attributed to the Lower Cambrian (Kotas 1973b), on the basis of lithological similarities with strata in the Goczałkowice IG-1 borehole containing a trilobite fauna and comparable trace fossils. On the basis of preliminary acritarch studies, parts of the succession were considered to be Middle Cambrian and Upper Cambrian – ?Tremadocian (Kowalczewski *et al.* 1984). Kowalczewski *et al.* (1984) and Kowalczewski (1990) interpreted the microfossil record as suggesting a partly inverted and repeated succession at Sosnowiec. This suggestion was subsequently questioned by Moczydłowska (1993a), who presented an alternative interpretation of the micropalaeontological data. The acritarch succession and the relative ages indicate a normal stratigraphic succession of strata underlying the sub-Devonian unconformity (Moczydłowska 1993a, b, 1995a, 1996a). The present investigation provides more accurate evidence on the relative ages and confirms that the succession spans the Middle–Upper Cambrian (see under 'Acritarch-based biochronology').

The Goczałkowice succession

The Cambrian sedimentary succession in the Goczałkowice IG-1 borehole is referred to the informal Goczałkowice formation (Kotas 1973a, 1982b; Fig. 8 herein). It occurs below the regional unconformity at the base of the Old Red Sandstone and above intrusive igneous rocks that directly overlie a folded and metamorphosed basement sequence (Cebulak *et al.* 1973a, Kotas 1973a, Cebulak & Kotas 1982). The Goczałkowice formation has a low angle of dip (5–16°) and is unmetamorphosed. The state of lithification is much less than in the Sosnowiec succession, and this is clearly evident from the preservation of the trace fossils. Trace fossils are not much affected by secondary cementation and are easy to separate from the rock (Fig. 9). The Goczałkowice formation consists predominantly of mudstones, alternating mudstones and fine-grained sandstones, and haematitic polymictic sandstones with conglomerates at their base. The total vertical thickness of the succession is 364 m (without dip correction). It overlies with pronounced angular unconformity rocks of the Precambrian basement, although the contact is displaced by the subsequent intrusion of a magmatic sill (Cebulak & Kotas 1982). The Precambrian sequence is referred to the informal Czéchowice formation (Kowalczewski 1990).

The lowermost part of the Goczałkowice formation is formed by polymictic conglomerates consisting of well-rounded quartz and quartzite pebbles in a polymodal quartzitic matrix at a depth of 3110.5–3129.2 m (Kotas 1973a, 1982b). The rocks are strongly haematized, being red-brown in colour, but become greyish-green at their base due to chloritization at the contact with the igneous

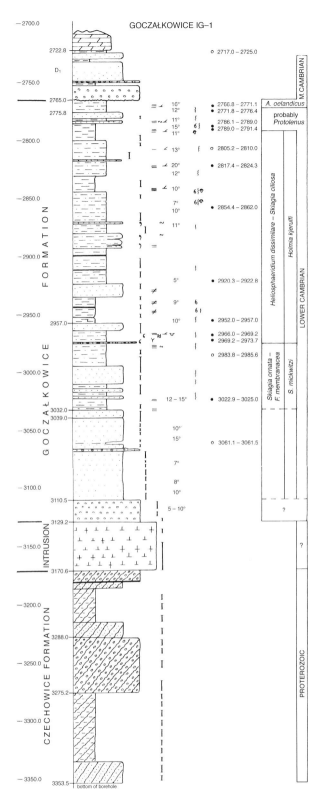

Fig. 8. Stratigraphic section of the Proterozoic basement complex and the Lower–Middle Cambrian succession in the Goczałkowice IG-1 borehole. Lithostratigraphic subdivision according to Kotas (1973a) and Kowalczewski (1990). Additional observations on sedimentary structures and occurrence of trace fossils added here. Occurrence of trilobites and brachiopods according to Biernat & Baliński (1973), Biernat *et al.* (1973), Kotas (1973a) and Orłowski (1975). P. Mb. = Pszczyna member, D1 = Lower Devonian. Graphic symbols as in Fig. 6.

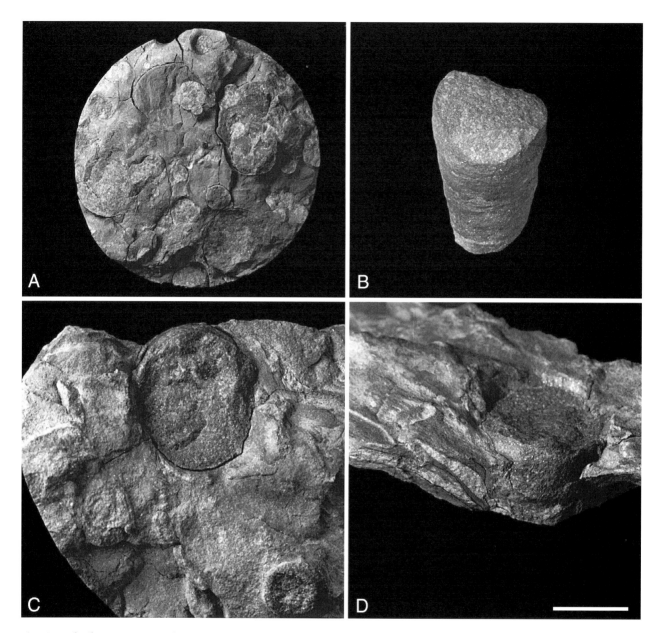

Fig. 9. Trace fossils, *Bergaueria* sp. and other simple burrows, from the Goczałkowice IG-1 borehole, interval 2973.7–2980.3 m, Goczałkowice formation, Lower Cambrian, time-equivalent to the *Schmidtiellus mickwitzi* Zone. □A. Fragment of drillcore (10 cm in diameter) showing the upper bedding surface with various cross-sections of dwelling burrows vertically penetrating the sediment. Diameter of burrows 0.4–2.8 cm. PMU-Pl.165. □B. *Bergaueria* sp., burrow cast separated from host rock. Diameter in the upper portion 2.7 cm, length of specimen 3.5 cm. PMU-Pl.166. □C–D. Fragment of core with *Bergaueria* sp. PMU-Pl.167. □C. The upper surface of bedding plane with cross-section of the specimen (2 cm in diameter) and additional small burrows. □D. Lateral view of the same specimen showing distortion of the primary stratification of alternating sandstone (light colour) and mudstone (dark grey) layers around burrow. Scale bar in D equals 3.2 cm for A; 1.7 cm for B; 1.3 cm for C and D.

rocks. An indistinct bedding observed in this interval displays a 5–10° dipping (Kotas 1973a). These conglomerates are 18.7 m thick and belong to the Pszczyna member (Kowalczewski 1990), which grades into polymodal haematitic quartz sandstones with thin conglomerate intercalations and occasional mudstones at a depth of 3032.0–3110.5 m. The sandstones, attaining a thickness of 78.5 m, are poorly sorted, mostly coarse-grained in the basal portion, and gradually fining upwards. Parallel bed-

ding, grading and muddy laminae have been observed in the upper part of these strata. A distinctive reddish conglomerate and conglomeratic sandstone bed occurs between 3066.1 and 3068.0 m (Fig. 8). The haematitic sandstones at depth of 3039.0–3110.0 m, were attributed to the informal '*Skolithos* member' (Kotas 1973a, 1982b; Kowalczewski 1990). However, no *Skolithos*-type nor any other trace fossils were observed during the course of the present study of the drillcore, and ichnofossils were not recorded in the borehole log. The sharp lithological boundary between the reddish quartz sandstones and the overlying sequence of predominantly greyish and bioturbated mudstones occurs at a depth of 3032.0 m (personal observation, depth according to borehole log). Mudstones, alternating mudstones and fine-grained quartz sandstone, and minor beds of pure sandstone, about 267 m thick (without correction for dip), comprise the upper part of the formation. These deposits are well-bedded and parallel-bedded, only a thin interval being massive and poorly stratified. Small-scale cross-bedding, wave lamination and abundant bioturbation are frequent within the strata. Glauconitic sandstones form a few minor thin interbeds in the upper part of the formation. A thin conglomerate layer with pebbles of phosphatic mudstone occurs in the lower part of this sequence (2973.5–2973.7 m; Kotas 1973a), reflecting a shallowing episode. Synaeresis cracks and load casts were also observed, during the present examination, in the overlying mudstones with phosphatic nodules. Trace fossils are abundant and well-preserved in the immediately underlying mudstone, with intercalations of fine-grained quartzitic sandstone (2973.7–2980.3 m). They display complex tiers of vertical burrows of variable diameter produced by benthic suspension feeders, e.g., *Bergaueria* sp. and others (Fig. 9). Early Cambrian trilobites indicating a *Holmia* age and inarticulate brachiopods were recorded at various depths (Kotas 1973a; Orłowski 1975; Fig. 8 herein). Based on a preliminary report on acritarchs (Moczydłowska *in* Kowalczewski *et al.* 1984), the succession was suggested to span the Lower to Middle Cambrian. This is confirmed by the present study. However, the previously mentioned occurrence of microfossils at a depth of 3177.6–3180.2 m (Moczydłowska *in* Kowalczewski *et al.* 1984, Kowalczewski 1990) was incorrect, being the result of an unfortunate sampling error. Re-examination of the drillcore revealed that this core interval consists of metamorphosed and mylonitised rocks only. The present record of acritarchs provides additional constraints on the relative age of the succession (see under 'Acritarch-based biochronology').

The Goczałkowice formation was interpreted as a transgressive detrital succession that accumulated in various environments (Kotas 1973a, 1982b). The basal portion (3039.0–3129.2 m) was inferred to have been deposited in tidal and coastal beach zones, the latter with possible alluvial influx. Succeeding rocks were interpreted to have accumulated in a tidal zone (interval of 2957.0–3039.0 m) and in a deeper part of the basin below wave-base, possibly even by turbidity currents (2765.0–2957.0 m). The latter was based on an observation of the 'fractional bedding forming turbiditic rhythms' (Kotas 1973a, 1982b). However, the sedimentary structures do not conform with typical graded bedding cycles of turbiditic sequences, and the strata do not display any features char-

acteristic of turbiditic sedimentation (personal observation). This succession was deposited in an offshore environment in a shallow shelf, below normal wave-base, but probably not on a basin slope. This is indicated by the frequent small-scale cross-bedding and the presence of abundant vertical burrows and bioturbation in general (Fig. 8).

The Cambrian succession is truncated by an erosion surface and zone of weathering. It is unconformably overlain by Lower Devonian terrigenous rocks, succeeded by Middle Devonian dolostones (Kotas 1973a, 1982a). The junction is a 10° angular unconformity (Kotas 1973a, c). The Lower Devonian succession commences with basal conglomerates containing angular pebbles and conglomeratic sandstones overlain by quartzitic sandstone with interbeds of mudstone. These deposits are poorly sorted and poorly stratified and contain numerous erosive washout surfaces. They are interpreted as fluvial deposits (Kotas 1982b) and are here referred to the Old Red Sandstone facies. In the absence of fossils, the Devonian age of these strata and the overlying dolostones is inferred entirely from lithostratigraphical correlations (Kotas 1973a, 1982b).

An igneous sill, 41.4 m thick and consisting of gabbro-diabases and diabases, penetrates the base of the succession. Additionally, thin apophyses of diabases, less than 20 cm thick, occur within the sequence of mudstones stratigraphically higher in the succession (at depths of 2899.5, 2911.0 and 2944.7 m) and in the metamorphosed basement (3180.2 m) (Kotas 1973a; Cebulak & Kotas 1982). The diabases are petrographically and chemically similar to those within the sill. Igneous rocks are clearly differentiated within the major body of the intrusion, and their petrography and chemical composition indicate that they were derived from a gabbroic magma. Gabbros and gabbro-diabases occur in the central part of the intrusion, grading in the outer zones into augite diabases; hyalobasalts are present in the contact zones with the adjoining rocks (Cebulak *et al.* 1973a; Kotas 1973a). The igneous rocks have not been dated isotopically, so their age of emplacement is unknown.

The basement complex consists of anchimetamorphic rocks of the Czéchowice formation that are folded, have undergone cataclasis and are strongly mylonitised. Alternating greyish-green phyllites, metapelites, metapsammites and metaconglomerates, with a mylonitic breccia at the top, are part of the complex (Cebulak *et al.* 1973a, c; Kotas 1973a; Cebulak & Kotas 1982; Fig. 8 herein). The dip of the less deformed metapelitic and metapsammitic rocks in the lower portion of the formation is 45–70° (Kotas 1973a). A Proterozoic age for the metamorphosed basement complex is evident from the presence of Early Cambrian trilobites in the overlying sedimentary rocks (Kotas 1973a; Orłowski 1975; Cebulak & Kotas 1982).

The Potrójna succession

A gently dipping and unmetamorphosed siliciclastic succession occurs between the two major regional unconformities in the Potrójna IG-1 borehole, overlain by Devonian strata and underlain by a sequence of steeply dipping and weakly metamorphosed rocks consisting of metapelites and metapsammites (Ślączka 1975, 1976, 1985c; Bielewicz *et al.* 1985; Wieser 1985; Kowalczewski 1990; Fig. 10 herein). Two formations have been distinguished. The Potrójna Formation, formally recognized by Kowalczewski (1990), comprises haematitic polymictic conglomerates, 27.5 m thick (Ślączka 1985c). The overlying Jaszczurowa formation (informally proposed by Kowalczewski 1990) consists of quartzitic sandstone, with a total thickness of 158 m (without dip correction) (Ślączka 1985a, b). The basement complex is referred to the informal Skawa formation (Kowalczewski 1990). It is tectonically deformed, with beds dipping at 40–90°. Its rocks are metamorphosed into greenschist facies and were affected by cataclasis (Ślączka 1985a, b). It also comprises tectonic breccias and quartz and calcite veins (Ślączka 1976; Bielewicz *et al.* 1985). The succession consists of pelitic and psammitic rocks. The original rocks were probably red-brown siltstones and fine-grained quartzitic sandstones in the upper part, and alternating fine-grained to medium-grained quartzitic sandstones, mudstones and claystones, and fine-grained arkoses in the lower part (Ślączka 1976, 1985b; Bielewicz *et al.* 1985; Wieser 1985). The upper boundary of the Skawa formation is erosive (Ślączka 1976, 1985b; Bielewicz *et al.* 1985); the lower boundary has never been penetrated by boreholes. It has been attributed to the Proterozoic (Ślączka 1976, 1985a, b) or to the ?Vendian – ?Lower Cambrian (Kowalczewski 1990).

The Potrójna Formation overlies the metamorphosed sequence with a sharp angular unconformity. The haematitic polymict conglomerates of the formation have a low dip (15–20°) and contain generally well-rounded pebbles, dominantly fine to medium-sized, rarely coarse, and quartz grains in a sandy and muddy matrix. The pebbles comprise sandstone, shale and pelite, with rare igneous and metamorphic rocks (granitoids, biotite-schists and haematitic sericite schists) (Ślączka 1976, 1985a, b; Wieser 1985; Kowalczewski 1990). Conglomerates in the upper part of the formation are more massive and consolidated, containing more angular pebbles (Wieser 1985; Ślączka 1985a). Features common to fanglomerates are also present at some levels (Ślączka 1976, 1985b; Wieser 1985). Pebbles in the basal portion are discoidal in shape and imbricated (Ślączka 1976, 1986b; Kowalczewski 1990), and those comprising pelites and metamorphic rocks derived from the basement are more abundant (Ślączka 1985b). The top of the Potrójna Formation is located within an interval of strata having the same lithology and the same angle of dip of adjoining beds. The only observable change is the colour; red below the depth of 3466.0 m and grey above (Ślączka 1985a). These observations were made on well cuttings, and geophysical log data since this interval of strata has not been cored. Kowalczewski (1990) recognized the top of the formation at a slightly different depth of 3464.5 m based on geophysical data. The Potrójna Formation has been considered to be? Vendian (=?Eocambrian; Ślączka 1976, 1985a, b) or Lower Cambrian (Kowalczewski 1990). However, these estimations are circumstantial as there is no direct evidence to assess the age of the deposits.

The overlying Jaszczurowa formation is a succession of medium-grained to coarse-grained quartzitic and conglomeratic sandstone, with interbedded quartz conglomerates and an interval of alternating fine-grained to medium-grained sandstones and mudstones. It is light grey in colour throughout the succession and has a dip of 15–20° (Ślączka 1985a). The basal portion of the conglomeratic sandstones and sandstones is weakly consolidated (Kowalczewski 1990). The interbedded conglomerates and conglomeratic sandstones are cross-stratified and consist of angular or weakly rounded pebbles in sandy quartzitic matrix. Sandstones form thick packages with rare intercalations of mudstones. The sedimentary structures include small-scale cross-bedding and parallel bedding, less frequently herring-bone bedding. Bioturbation is neither abundant nor diverse, but occurs at various levels from a depth immediately above the lowermost conglomeratic bed in the formation (Ślączka 1985a; Fig. 10). It consists of simple vertical burrows of small diameter attributed to *Skolithos* and unidentified ichnotaxa. Load casts and small-scale slump structures occur in an interval of alternating sandstones and mudstones. In the uppermost part of the formation, glauconitic fine-grained sandstones occur (Ślączka 1985a). The upper boundary of the Jaszczurowa formation is erosional, coinciding with the pre-Devonian unconformity, and is marked by a sharp lithologic change. Though inconspicuous, the unconformable contact may be angular (Ślączka 1985b).

The Jaszczurowa formation was considered to be Early Cambrian on the grounds of lithological correlation with the succession in the Goczałkowice IG-1 borehole and the occurrence of similar trace fossils (Ślączka 1976, 1985a, b). Subsequently, it was correlated with the Middle and Upper Cambrian (Kowalczewski *et al.* 1984), based on discovery of acritarch taxa which range from the Middle Cambrian to Tremadoc between 3356.5 and 3363.5 m (Moczydłowska *in* Kowalczewski *et al.* 1984). Kowalczewski (1990) subsequently suggested that the lowermost part of the Jaszczurowa formation (the Mucharz member) is Late Cambrian in age. The additional evidence for the relative age estimation is provided by the present acritarch record supporting Late Cambrian age for all but the

Fig. 10. Stratigraphic section of the Proterozoic basement complex and unidentified Cambrian to Upper Cambrian succession in the Potrójna IG-1 borehole. Lithostratigraphic subdivision according to Ślączka (1985c) and Kowalczewski (1990). Occurrence of trace fossils according to Ślączka (1985c). D1=Lower Devonian, D2=Middle Devonian. For explanations see Fig. 6.

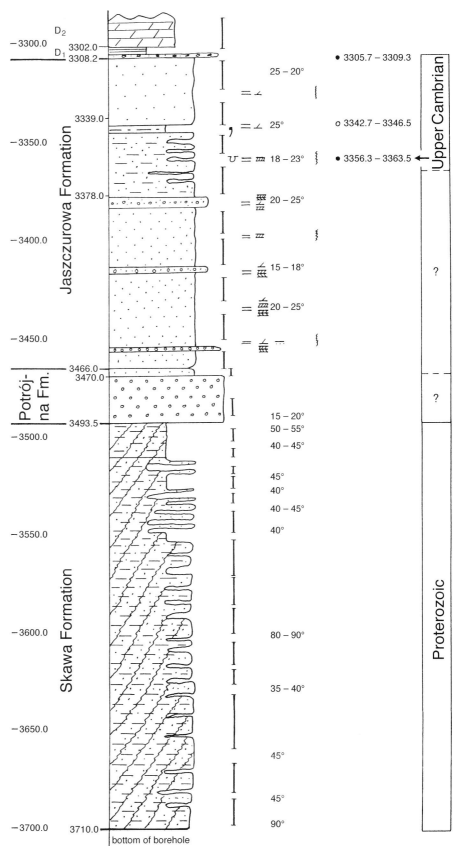

upper part of the Jaszczurowa formation (see under 'Acritarch-based biochronology').

The overlying, presumably Lower Devonian, deposits are conglomerates and sandstones, followed by chaotically textured claystones and shales, dipping at 20–25°. The rocks are unfossiliferous and regarded as continental, claystones in the sequence possibly representing palaeosols (Ślączka 1976, 1985b). The relative age of these strata is inferred from the lithostratigraphic correlation with beds consisting of Lower Devonian psilophyte remains (Ślączka 1985b). Carbonate rocks occurring stratigraphically higher in the succession are attributed to the Middle and Upper Devonian (Ślączka 1985b).

The entire succession of gently dipping and unmetamorphosed terrigenous rocks in the Potrójna IG-1 borehole has been regarded as representing a sedimentary cover of the Precambrian platform that has escaped any subsequent folding (Ślączka 1985b). The deposition of the Jaszczurowa formation, consisting of predominantly quartzitic sandstones, took place in a shallow marine environment within the tidal zone (Ślączka 1976, 1985b). The deposition of the haematitic polymictic conglomerates of the Potrójna Formation took place under continental conditions; the formation has been considered to represent Precambrian molasse deposits but erroneously referred to as the Assynthian phase of orogenic deformation (Ślączka 1976, 1982, 1985b; Kotas 1982a, Wieser 1985). The Assynthian tectonic event is attributed to the Late Cambrian – Tremadocian times (Roberts & Sturt 1980; Sturt *et al.* 1980; Brochwicz-Lewiński *et al.* 1981). The deposition of the Potrójna Formation predates Late Cambrian, as proven by the record of the Late Cambrian acritarchs in the overlying Jaszczurowa formation, and thus the formation cannot be referred to as the Assynthian molasse. It is here regarded as post-Cadomian molasse. This is inferred from the stratigraphic position of the formation between the tectonically deformed and metamorphosed Skawa Formation, evidently Precambrian in age, and the Upper Cambrian Jaszczurowa formation as documented by the acritarchs (see under 'Tectonic implications').

Taphonomy and palaeoenvironmental conditions

The state of preservation of the microfossils and their taphonomy is of particular interest to this study, since one of the aims is to establish whether investigated microfossils accumulated as a result of normal settling and burial (*in situ*), or as a result of possible sediment reworking. The biostratigraphy of the successions in Sosnowiec, Potrójna and Goczałkowice, in part devoid of trilobites, relies heavily on the microfossils, since these provide the only means for establishing the relative age of the investigated strata. Therefore the nature of the accumulation process of the microfossils is of utmost importance.

Microfossils recovered from the Sosnowiec succession are generally both abundant and well or satisfactorily preserved. A comparable state of preservation is observed among microfossils of various morphotypes and dimensions, including picoplanktic forms that are only a few micrometres in diameter. Most acritarchs from individual samples are preserved intact, with fragile ornaments and surface sculpture. However, there is always a number of specimens that display broken processes and corroded vesicle wall. Degradation of the vesicle wall as a result of the growth of pyrite (framboids and pseudomorphs) is more frequently observed among sphaeromorphs and species of *Comasphaeridium* (Fig. 22B, D), *Eliasum* (Fig. 28B, D), *Granomarginata* (Fig. 29D), *Multiplicisphaeridium* (Figs. 33A, C, E and 35) and *Solisphaeridium* (Fig. 43E, F). Corrosion of the wall, appearing as an uneven or roughly pseudogranular surface (Figs. 20I and 27I), particularly among leiosphaerids, could be confused with real primary sculpture of the vesicle (Fig. 27F, G). Despite preservation in fine-grained detrital rocks, three-dimensional preservation is very common at most stratigraphic levels, irrespective of the morphology of individual specimens. For instance, acanthomorphic species are exceptionally well-preserved in the intervals 3403.5–3407.2 m and 3174.0–3174.7 m (Figs. 30A–C, G–I; 33A–C, E–L; 43A–B), and discoidal specimens of *Granomarginata* (Fig. 29A) and herkomorphic *Cymatiosphaera* (Fig. 25B, C) at a depth of 3365.2–3372.3 m. It is significant that picoplanktic microfossils are recorded in all these samples. The assemblages comprise a comprehensive spectra of taxa probably representing favourable preservational conditions. Samples from the core intervals at 3419.6–3423.8 m and 3343.9–3347.6 m yielded less abundant and poorly preserved microfossils. The differences in the state of preservation are clearly related to the palaeoenvironmental conditions represented by these two sets of samples.

The siliciclastic succession at Sosnowiec was deposited in a shallow marine open-shelf basin, dominated by low-energy environments below wave base. Two depositional settings, offshore and nearshore, are inferred for various depth intervals in the borehole. Most of the succession consists of uniform strata of mudstones alternating with fine-grained sandstones and laminated claystones, well-bedded, frequently with parallel lamination and occasional small-scale cross-bedding. They are bioturbated and perpendicular and horizontal burrows are well-preserved at several horizons. This mudstone-dominated offshore environment is represented by samples deriving from depths of 3174.0–3174.7 m, 3210.0–3212.0 m, 3365.2–3372.3 m and 3403.5–3407.2 m. They yielded microfossil assemblages taxonomically diverse and fairly well-preserved (see above). Periodical shallowing of the shelf resulted in the deposition of fine-grained quartz sandstones with interbeds of mudstones and thin clayey intercalations in nearshore environments. These deposits, sampled at the depths of 3419.6–3423.8 m and 3343.9–3347.6 m, are poorly bedded and contain occasionally clay-draped scours and thin (a few centimetres thick) beds of graded-bedded sandstones and mudstones with erosional bases. They are also bioturbated and display predominantly vertical burrows. Microfossils preserved in these rocks are rare, being more often broken up and corroded, but their relative morphologic diversity, taking into consideration the total number of specimens within the assemblages, is not much different from those described above. This suggests that the spectrum of taxa and abundance of specimens preserved in the near-shore environments is controlled by the preservational factors, rather than by strictly palaeoecological control mechanisms. In such environments the preservation potential of microfossils is strongly affected by mechanical degradation of microorganisms after death, by turbulence and friction between the detrital sediment particles in high-energy conditions, and by the activity of the benthic animals, strongly bioturbating the bottom sediments.

The thermal alteration of the organic matter, as shown by the colour of acritarchs, is more or less uniform throughout the whole succession. The colour varies from honey yellow to light brown and brown. Subtle differences are observable between thick and thin-walled vesicles, and between specimens having hollow or solid ornament (processes), but are unrelated to the size of microfossils. The particulate organic matter, very abundant in the Sosnowiec succession, is much darker (dark brown to black in colour). The colour of acritarchs corresponds to the values of the thermal alteration index (TAI) 2–3, according to Hayes *et al.* (1983), or 4–5 according to the scale of Rovnina (1981). The inferred palaeotemperatures to which the deposits were exposed during their geological history are in the range of 100–150°C or 120–180°C, respectively. The stage of lithogenesis of the strata might, in any case, be referred to mesocatagenesis, thus not exceeding a palaeotemperature of 180°C and pressures of 1 kbar (Rovnina 1981; Hayes *et al.* 1983). Neither facies relationships, sedimentary structures nor stages of thermal maturation of the organic matter could be interpreted to suggest reworking and reincorporation of acritarchs in the sediments in the Sosnowiec succession. This effectively rules out any suspicion of stratigraphically mixed fossil assemblages.

Microfossils recorded in the Goczałkowice succession are not numerous but are fairly well-preserved. The variability of morphotypes is high (14 recorded taxa) in relation to the small number of preserved specimens. The size

distribution is homogenous, most specimens being within the interval of 20–40 μm, but minute vesicles less than 10 μm in diameter are also present. The thermal alteration of the organic matter is significantly lower than in the Sosnowiec assemblages. The colour of acritarchs ranges from yellow to dark yellow, consistent with TAI 1–2 (Hayes *et al.* 1983), corresponding roughly to a TAI 2–3 of Rovnina (1981). The corresponding palaeotemperatures inferred from these values are 50–100°C or 75–120°C, respectively. The thermal maturation of the organic matter and inferred palaeotemperatures are in accordance with the early stage of protocatagenesis in the process of lithogenesis, implying also low pressures (Rovnina 1981). All fossiliferous samples (Fig. 8) are from mudstones and mudstones interbedded with very fine-grained sandstones, forming a monotonous sequence exceeding 200 m in thickness. The deposits are well bedded or locally laminated, with small-scale cross-bedding and wave lamination occurring at several levels. The entire sequence is bioturbated. Trilobite fragments occur at two intervals, and brachiopod moulds were recorded at several horizons. The facies association of this sequence is inferred to have developed in an offshore setting of a siliciclastic shallow shelf. The state of preservation of microfossils and their thermal alteration and distribution within the host rocks is in accordance with the sedimentary development and the stage of lithogenesis of the deposits observed in the succession. It is therefore concluded that acritarchs at Goczałkowice occur in a primary burial setting (*in situ*).

The acritarchs recovered in the Potrójna succession are scarce and poorly preserved. Ornamented specimens are frequently broken and sphaeromorphs are corroded. The colour of acritarchs is yellow to dark yellow. The particulate kerogen, light brown to brown in colour, is dispersed. Observations on the thermal alteration are limited, but the stage of maturation of the organic matter is similar to that in the Goczałkowice borehole. The fossiliferous interval at 3356.5–3363.5 m (Fig. 10) consists of fine-grained to medium-grained quartz sandstones with thin interbedded mudstone displaying various bedding structures, small-scale cross-bedding, small-scale slumping structures, load casts and vertical bioturbation burrows. The formation was deposited in near-shore, tidally influenced environments. Short-distance transport of microfossils within the tidal flat zone in this very shallow part of the basin is possible. However, the possibility of redeposition of microfossils from stratigraphically older deposits is excluded, because that sequence in question overlies metamorphosed basement rocks in this area. Reworking from longer distances is not likely, because acanthomorphic specimens are preserved with long processes which would be prone to destruction.

Micropalaeontological record

Organic-walled microfossils recorded in Upper Silesia are taxonomically diverse and attributed to 75 form-species, but they are not particularly abundant. More than 1700 diagnostically ornamented specimens were identified, and hundreds of sphaeromorphic microfossils were also recorded. The latter category comprises specimens that differ morphologically only with respect to the overall diameter of the vesicle and to wall thickness. These microfossils are attributed to *Leiosphaeridia* sp. and are here treated as a single 'form-taxon'. In my opinion, the recognition of species based on size clusters alone is untenable and results in artificially high biodiversity estimates that are not supported by sounder taxonomic criteria.

All microfossils recovered in this study are of Cambrian age and are referred to various biochrons or, more generally, to epochs. The taxa occurring at each individual stratigraphic level are further referred to as assemblages. Assemblages belonging to the same geochronologic/chronostratigraphic unit (age, epoch, period/stage, series, system) are regarded as an association (e.g., Cambrian association, Lower Cambrian association etc.).Compared to the total number of Cambrian taxa (333), the taxonomic diversity of microfossils from Upper Silesia is relatively low and might be estimated as 22% of all known species.

Most acritarch species have well-recognized stratigraphic ranges, enabling the relative ages of the assemblages to be estimated (see under 'Acritarch-based biochronology'), and their geographic distribution used to infer possible palaeogeographic relationships (see under 'Palaeobiogeography').

In the Cambrian acritarch associations from Upper Silesia, 18 taxa are described as new species and 6 taxa are left under open nomenclature. Three new taxa were formerly described under open nomenclature or erroneous taxonomic epithets, and are here formally transferred to new species. These species are already known from a number of occurrences. The new taxa and those left under open nomenclature constitute a substantial part of the entire association (about 27.6%) and, although not used for biostratigraphy or palaeogeography, they are significant from the point of view of palaeobiology and biodiversity. The new species are distinctly ornamented, and most of them are quite abundant in the present material.

The majority of microfossils from Upper Silesia (32 taxa corresponding to 42% of the whole association) are cosmopolitan species. Here acritarch species are considered cosmopolitan whenever their palaeogeographic distributions are known from at least three discrete palaeocontinents recognized during Cambrian times. The identification of Cambrian palaeocontinents and the terranes that were incorporated with these continents into the same plates during Cambrian times follows McKer-

row & Cocks (1995). The location of the Upper Silesia terrane within East Avalonia follows Moczydłowska (1995b, 1996a, 1997). Cambrian palaeocontinents with up-to-date records of acritarchs comprise Baltica, Laurentia, Siberia and Gondwana (with a marginal assembly of microplates referred to Avalonia, Armorica, South China and Australia). Among cosmopolitan acritarch taxa are various species of *Skiagia*, *Heliosphaeridium*, *Asteridium*, *Polygonium*, *Lophosphaeridium*, *Multiplicisphaeridium* and *Revinotesta*. The additional 14 species within the Cambrian association from Upper Silesia (18.4% of the association) have widespread palaeogeographic distributions. A widespread distribution is here inferred for species occurring contemporaneously in the shelves of two palaeocontinents. Such species belong to *Cristallinium*, *Timofeevia*, *Eliasum* and *Multiplicisphaeridium*. Nine species (11.8% of the association) appear to be sparse or endemic and are recorded only in the shelf of one palaeocontinent. Species such as *Duplisphaera luminosa* n.comb., and *Adara undulata* n.sp. (previous *nomina nuda*), and *Solisphaeridium flexipilosum* emend. were thus far recorded only in Armorica (Iberia and Bohemia) and are presently known in East Avalonia (Upper Silesia).

The record of microfossils from Upper Silesia, in which over 60% of species are cosmopolitan and widely occurring, provides additional evidence that there were no restricted acritarch bioprovinces during Cambrian times. This was previously suggested for the Early Cambrian (Moczydłowska & Vidal 1992) and is herein extended to the Middle Cambrian. The worldwide distribution of numerous planktic taxa (see compilation under 'Palaeontological descriptions') indicates the absence of palaeoenvironmental barriers, thus allowing free plankton dispersal between the shelves of various palaeocontinents (see under 'Palaeobiogeography').

Microfossils with extremely small overall dimensions are here referred to as picoplankton. The category of picoplankton is applied to Recent microorganisms with cell size less than 2 μm diameter, predominantly prokaryotes but also photosynthetic eukaryotes (Courties *et al.* 1994). Here, this denomination is extended to microfossils of slightly larger dimensions (around 4–6 μm in diameter). Acritarchs with such dimensions were previously described from the Mesozoic and Cenozoic (Habib 1972, 1979; Habib & Knapp 1982), and from Cambrian strata (Vanguestaine 1974; Jankauskas 1975; Moczydłowska 1988, 1991). They were informally called 'small acritarchs', referring to their dimensions generally less than 10 μm (Habib & Knapp 1982). Because of the difficulties in studying fossil picoplankters, their taxonomy and potential use in biostratigraphy was limited. However, scanning electron microscopic studies of some of these microfossils have revealed the presence of diagnostic vesicle and process ornamentation that allow recognition of discrete genera and species (Habib & Knapp 1982),

Fig. 11. Acritach radiation and extinction rates during Cambrian times. Global record of Cambrian species and their stratigraphic ranges compiled from various sources. LC=Lower Cambrian, Plat.=*Platysolenites* Biochron, Schm.=*Schmidtiellus* Biochron, Holm.=*Holmia* Biochron, prot.=*Protolenus* Biochron, MC=Middle Cambrian, oel.=*Acadoparadoxides oelandicus* Biochron, parad.=*Paradoxides paradoxissimus* Biochron, forch.=*Paradoxides forchhammeri* Biochron, UC=Upper Cambrian, Agn.=*Agnostus* Biochron, Olen.=*Olenus* Biochron, Parab.=*Parabolina* Biochron, Lept.=*Leptoplastus* Biochron, Pelt.=*Peltura* Biochron, Acer.=*Acerocare* Biochron. Modified after Vidal & Moczydłowska (1995b). Based on data base of the global records compiled from various sources.

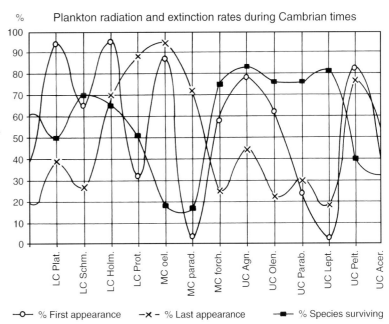

thought this is not accepted by some workers (e.g., Sarjeant & Stancliffe 1994). Numerous picoplanktic microfossils from Upper Silesia are attributed to *Asteridium* and *Revinotesta*. They co-occur within assemblages consisting of a wide array of morphotypes and dimensions, including, for example, large specimens of *Eliasum* and *Cristallinium*. Picoplankton do not appear to occur together with any particular taxonomic or dimensional class of microfossils. The taxonomic composition of each investigated sample appears to be random, and in the studied sections it seems to be subjected only to stratigraphically controlled change.

A detailed analysis of the present Cambrian acritarch association from Upper Silesia provides observations on the taxonomy of microfossils and constrains the biochronology of the investigated strata. The stratigraphic succession of age-diagnostic assemblages and the taphonomy of microfossils, added to observations on the sedimentary succession and state of preservation of trace fossils, has implications for hypotheses dealing with the tectonic development of the area.

Biodiversity of Cambrian phytoplankton

The Cambrian Period is marked by profound and unprecedented biotic diversification referred as to the 'Cambrian explosion' (Conway Morris 1987, 1989, 1992; Valentine 1994) or 'Cambrian revolution' (Seilacher 1994). This event is evidenced by the appearance of most modern metazoan phyla (Valentine 1992; Conway Morris 1994; Bergström 1989, 1994; Bengtson & Conway Morris

1994), and is also accompanied by an exponential radiation among photoautotrophic protoctists (Volkova *et al.* 1979; Vidal & Knoll 1983; Vidal 1984; Moczydłowska 1991; Vidal & Moczydłowska 1992) and thallophytes (Hofmann 1994; Sun Weiguo 1994; Riding 1994; Butterfield *et al.* 1994; Steiner 1994).The biodiversity of acritarchs is of a particular interest because, being interpreted as primary producers, they probably affected the entire food web. Acritarchs have a comprehensive record of evolution in Proterozoic (Mendelson & Schopf 1992; Knoll 1994a; Vidal 1994) and early Phanerozoic times (Tappan 1980; Martin 1993). However, any appreciation of Cambrian phytoplanktic diversity is tentative because it demands taxonomic revisions and reassessment of the stratigraphic ranges of some taxa in relation to a global geochronologic frame. Despite a biased fossil record (Vidal 1994) and difficulties in assessing the true number of fossil taxa owing to the taxonomic inconsistencies, the recognizable changes in phytoplankton associations between discrete biochrons reflect major biotic events such as radiations, decline in diversity, and extinctions (Fig. 11). In any evaluation of possible biodiversity trends using micropalaeontological data, it is evident that discrete biochrons are unequally represented. Furthermore, individual biochrons span time intervals of differing length. The pattern of Cambrian secular biodiversity changes (Knoll 1994b; Vidal & Moczydłowska 1995, 1997) is well recognized in the acritarch record from Upper Silesia and is further elucidated (Fig. 12). The acritarch record from Upper Silesia comprises most of the Lower and Middle Cambrian, which is represented by fossiliferous sedimentary successions in which only the basal

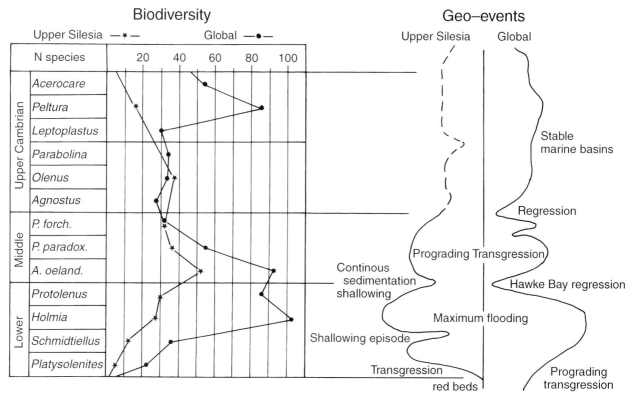

Fig. 12. Acritarch biodiversity and geo-events in the Upper Silesia and global records during Cambrian times. Global data compiled from various sources and this study. Data base available on request (Vidal & Moczydłowska 1997) .

portion of the Lower Cambrian has not been recorded, though it may be represented by unfossiliferous red beds in the Goczałkowice borehole. The Upper Cambrian is only partly represented and imprecisely identified in the Sosnowiec and Potrójna boreholes. The existing portions of the Lower and Middle Cambrian succession are comprehensively developed and may contain only minor sedimentary breaks which do not seem to reduce significantly the recognized succession of biozones (Figs. 6 and 8). The investigated strata accumulated in nearshore and offshore open-shelf settings of the depositional basin that was connected, as inferred from the palaeobiogeographic evidence, with extensive shelf areas expanding along and/ or between the Cambrian palaeocontinents (see under 'Palaeobiogeography'). General biodiversity trends in Upper Silesia and in associations reported from coeval strata in Baltica (Volkova *et al.* 1979; Hagenfeldt 1989b; Moczydłowska 1991), Laurentia (Vidal & Peel 1993) and West and East Avalonia (Martin & Dean 1983, 1988; Martin *in* Young *et al.* 1994) are similar. Thus, it is here inferred that the biotic record from Upper Silesia may reflect global fluctuations in phytoplankton diversity during Cambrian times. There are, however, some slight discrepancies in diversity patterns that are interpreted as indicating regional features. Such regional variations are detected at the Lower–Middle Cambrian boundary.

Fluctuations in the total number of species and the replacement of taxa seem to relate globally to contemporaneous geological events. Bio-events appear to parallel or closely follow geo-events with some time lapse, thus suggesting that most of the recorded biotic changes in Cambrian times may have been induced or influenced by palaeoenvironmental causes. Although the fossil record allows substantial changes in fossil populations to be discerned, the nature of the biotic innovations that might have triggered the Cambrian radiations remains unclear. The relationships between biological and environmental interactions (mostly through nutrient recycling, oxygen level and temperature changes) and their impact on evolutionary changes are not fully understood. It has been hypothesized that Proterozoic and Cambrian tectonic events and geochemical evolution ushered in environmental changes that prompted biological innovations (e.g., the emergence of metazoans; Cloud 1968a, b, 1976, 1988; Brasier 1991, 1992; Knoll 1992a; Knoll & Holland 1995). On the other hand, physicochemical events may not have been linked to the factors that induced biotic change (Knoll 1994b). Hence, it cannot be discounted that the radiations of the Cambrian may have had biological causes alone. This was suggested for the major protist diversification during Proterozoic times which is not associated with any evident environmental change (Knoll

1994b). However, the timing of geo-events in the Cambrian has much better resolution than in the Proterozoic, and this allows biological changes to be related more precisely to geological phenomena. The record of contemporaneous Cambrian geo-events and bio-events (Fig. 12) occurring in short time intervals of only a few million years (see under 'Geochronologic duration'), provides additional evidence that many of the observed evolutionary changes were probably triggered environmentally. However, the question of how far environmental change has affected evolutionary innovations (Cloud 1968a, b, 1976; Brasier 1991, 1992; Knoll 1992a; Knoll & Holland 1995), or to what extent the biosphere has affected past environments (Lovelock & Margulis 1974; Lovelock 1979, 1992; Margulis & Olendzenski 1992; Knoll 1992b, 1994b), remains undetermined.

During Early Cambrian times, acritarch diversity gradually increased, following the rapid radiation at the Precambrian–Cambrian boundary. This trend extended into the *Platysolenites* Biochron (Moczydłowska & Vidal 1986; Moczydłowska 1991; Vidal & Moczydłowska 1995) and continued throughout the *Schmidtiellus* Biochron, a period of growing shelf basins rich in nutrients and well-oxygenated surface layers. The conditions were those of a prograding transgression and sea-level rise (Brasier 1980, 1990, 1991). The highest taxonomic diversity and numerical abundance of phytoplankton is observed within the *Holmia kjerulfi* Biozone (Fig. 12), which represents an acme in Early Cambrian phytoplankton speciation. Interestingly, this coincides with maximum flooding of the Early Cambrian transgression (Bergström & Gee 1985; Mens *et al.* 1990; Link 1995). Subsequently, acritarch diversity declined markedly in the *Protolenus* Biozone, because of the initiation of a severe extinction event, with a synchronous moderate rate of surviving species and a very low rate of speciation. However, a substantial number of species survived (around 50% of the global record, Fig. 11; Vidal & Moczydłowska 1997), as shown by the extension of their ranges across the Lower–Middle Cambrian boundary (Volkova *et al.* 1979; Martin & Dean 1983; Martin *in* Young *et al.* 1994; Hagenfeldt 1994; Fig. 14). In Upper Silesia, the micropalaeontological record shows that most of the Early Cambrian acritarch species crossed this boundary (Fig. 16). This low diversity of phytoplankton in the *Protolenus* Biozone coincides with a shallowing event that ended in some areas with the Hawke Bay regression (Palmer & James 1980; Bergström & Gee 1985; Vidal & Moczydłowska 1996). The Hawke Bay Event is reflected by the hiatus separating Lower and Middle Cambrian successions in North America (Palmer & James 1980). It has also been recognized in some areas of Scandinavia, in Norway and Sweden, and has been suggested to be of a global significance (Bergström & Gee 1985). The observed discrepancy between acritarch diversity in Upper Silesia (showing a progressive increase in biodiversity) and the cumulative global diversity (extinc-

tion and new radiation, resulting in an almost constant level of diversity; Fig. 12), may be explained in terms of variations reflecting regional environmental conditions, not being affected by inferred global geo-events.

A significant acritarch radiation is documented in the *Acadoparadoxides oelandicus* Biozone. The percentage of newly appearing species in the acritarch association of this time interval was higher than at any other time in the Cambrian (Fig. 11). This radiation was of the same magnitude as those of the *Platysolenites* and *Holmia kjerulfi* Biozones, but contrary to them it was preceded by and was contemporaneous with the two-staged extinction event. The two-staged extinction began in the *Protolenus* Biozone and continued in the *A. oelandicus* Biozone. The first stage (*Protolenus*) was less drastic because it was paralleled by a moderate rate of surviving species (see above), while the second stage (*A. oelandicus*) was more severe as it was characterized by the highest rate of extinction and the lowest rate of surviving species in the entire Cambrian Period. This extinction is regarded as two-staged event rather than two individual extinctions because of the estimated short time interval separating both episodes.

Considering all changes in diversity, the *A. oelandicus* Biochron seems to be the time of the most significant turnover in phytoplankton during the Cambrian. This happened during the transgression indicated by overstepping *A. oelandicus* strata (Lendzion 1983a), which continued uninterrupted from Early Cambrian times, as evidenced by numerous sedimentologically continuous successions across Lower–Middle Cambrian boundary worldwide (Brasier 1989b; Mens *et al.* 1990; Landing 1992; Liñán *et al.* 1993b; Ineson *et al.* 1994). The ensuing Middle Cambrian times were characterized by the gradual decline and then increase in diversity of phytoplankton in the global records (Figs. 11 and 12). The extinction rate was still relatively high in the *P. paradoxissimus* Biochron, paralleled by moderate rate of surviving and very low rate of new-appearing species. This decline was recovered and stabilized soon after, in the *P. forchhammeri* Biochron. In the latter time interval the high rates of surviving and new-appearing species surpassed the low rate of extinction, resulting in a significant increase of biodiversity. The observed fluctuations in acritarch diversity during the Middle Cambrian are recorded in environments established in the stable marine basins, with transgression and regression episodes related to the local instabilities and eustatic sea level changes caused by the spreading of the Iapetus Ocean (Bergström & Gee 1985). The sedimentary successions that accumulated in the shelf and slope conditions were often condensed kerogen-rich shale and minor carbonate sequences (e.g., Alum Shale in Scandinavia, Gee 1972) or carbonate-starved argillites (Laurentia, Ineson *et al.* 1994).

Similar, mostly stable geotectonic conditions extended into Late Cambrian times. Condensed black shale sequences or fine-grained siliciclastic rocks accumulated on the passive margins of Baltica (Bergström & Gee 1985; Welsch 1986; Mens *et al.* 1990). Fine-grained and coarse-grained siliciclastic sequences were deposited in the destructive margins of Gondwana (NE Ireland, Crimes & Crossley 1968, Crimes *et al.* 1995; Wales, Young *et al.* 1994; SE Newfoundland, Martin & Dean 1988; Iberia, Aramburu *et al.* 1992; Morocco, Geyer & Landing 1995) or carbonate platform sequences in its inner shelf areas (continuous throughout Lower–Upper Cambrian in Precordillera of Argentina, Baldis & Bordonaro 1985, Bordonaro 1992). The Middle–Upper Cambrian boundary is sedimentologically continuous in various areas in Baltica (NW Poland, Russia, Sweden, Norway Finnmark; Mens *et al.* 1990; Welsch 1986), Gondwana (Precordillera in Argentina, Bordonaro 1992), and Laurentia (Greenland, Ineson *et al.* 1994). In the Late Cambrian Epoch two significant events affecting the phytoplankton diversity have been observed (Fig. 11). These are the radiation in the *Agnostus* Biozone – which together with a high rate of surviving species gradually increased the acritarch biodiversity at the time – and the turnover in the taxonomic composition of acritarch associations observed in the *Peltura* Biozone. The latter event is depicted by the high rate of extinction and the even higher rate of contemporaneous radiation along with a substantial number of surviving species. During the intermediate times in the Late Cambrian, the stasis in biodiversity is inferred from the record of high rate of surviving species paralleled by low rate of extinction and very low rate of new appearing species.

In summary, the major bio-events recorded by phytoplankton diversity in Cambrian are:

- major radiations at the Precambrian–Cambrian boundary and in the *Platysolenites* Biochron (with very few surviving species from the pre-existing depauperate assemblage), and in the Early Cambrian *Holmia kjerulfi* Biochron (high rate of radiation enhanced by high rate of surviving species and low extinction),

- less significant radiation in the Late Cambrian *Agnostus pisiformis* Biochron (with additionally high surviving rate and low rate of extinction),

- major turnovers of phytoplankton in the Middle Cambrian *A. oelandicus* Biochron and in the Late Cambrian *Peltura* Biochron (the highest rate of extinction and the highest rate of radiation),

- the two-staged extinction in the Early–Middle Cambrian transition, during the *Protolenus* and *A. oelandicus* Biochrons,

- minor extinction in the Middle Cambrian *P. paradoxissimus* Biochron (with low rate of surviving and lower rate of new appearing taxa).

Geochronologic duration of Early Cambrian zones

The duration of Early Cambrian acritarch biochrons and their time-equivalent trilobite biochrons has been estimated to 4–5 Ma (Vidal *et al.* 1995) or 2–4 Ma (Compston *et al.* 1995). This was based on isotopic age determinations of rock units either comprising age-diagnostic faunal and acritarch assemblages or being part of successions containing such assemblages. The absolute age obtained for strata near the base of the Cambrian in Siberia is 543.9±0.2 Ma, whereas a maximum age of 534.6±0.5 Ma was obtained for the Middle Tommotian *Dokidocyathus regularis* Zone (Bowring *et al.* 1993; biochronologically re-evaluated by Vidal *et al.* 1995 and Compston *et al.* 1995 concerning the Tommotian age). However, because of uncertainties in the recognition of the base of Cambrian in Siberia and its correlation with the world stratotype section in Newfoundland, the proposed age of ca. 544 Ma for the beginning of Cambrian (Bowring *et al.* 1993) might refer to Neoproterozoic strata (Vidal *et al.* 1995). The time span of 8.6–10.0 Ma calculated from the above ages embraces two acritarch zones, the *Asteridium tornatum* – *Comasphaeridium velvetum* and *Skiagia ornata* – *Fimbriaglomerella membranacea* Assemblage Zones, which correspond to the *Platysolenites antiquissimus* and *Schmidtiellus mickwitzi* faunal Zones, respectively (Moczydłowska 1991; Vidal *et al.* 1995). However, if the age of 544 Ma is proven to be latest Neoproterozoic, this interval would become shorter. The recently obtained age for the Precambrian–Cambrian boundary in Namibia is constrained within the range 539.4±1 Ma to 543.3±1 Ma (Grotzinger *et al.* 1995). Depending on whether the minimum or maximum age is accepted as close to the age of the boundary, the time interval of about 4–10 Ma would apply for the two biochrons in question (Fig. 13). This gives minimum and maximum durations of 2–5 Ma per biochron. The assumption of near equal duration of the two biochrons is supported by the comparable thicknesses of the strata referred to these zones in the sedimentologically homogenous and continuous siliciclastic successions in the type area (Moczydłowska 1991).

The age 534.6±0.5 Ma for the Middle Tommotian *Dokidocyathus regularis* Zone in Siberia (Bowring *et al.* 1993; re-evaluated by Vidal *et al.* 1995; Compston *et al.* 1995) is biochronologically constrained by the occurrence of the age-diagnostic fossils overlying the dated strata (Repina *et al.* 1974; Rudavskaya & Vasileva 1984; Vasileva 1985; Vidal *et al.* 1995). Accepting this age and the palaeontologic evidence, the isotopic date of 530.7±0.9 Ma for the middle part of the *Rusophycus avalonensis* Zone in New Brunswick, regarded as a sub-Tommotian (or Manykaian) equivalent (Isachsen *et al.* 1994), is biostratigraphically inconsistent (Vidal *et al.* 1995). The dated strata are

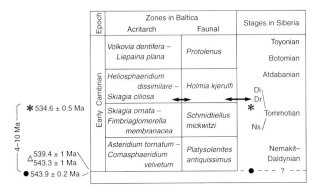

Epoch	Zones in Baltica		Stages in Siberia
	Acritarch	Faunal	
Early Cambrian	*Volkovia dentifera – Liepaina plana*	*Protolenus*	Toyonian
			Botomian
	Heliosphaeridium dissimilare – Skiagia ciliosa	*Holmia kjerulfi*	Atdabanian
	Skiagia ornata – Fimbriaglomerella membranacea	*Schmidtiellus mickwitzi*	Tommotian
	Asteridium tornatum – Comasphaeridium velvetum	*Platysolenites antiquissimus*	Nemakit–Daldynian

Dl, Dr, Ns = zones in Tommotian

✳ 534.6 ± 0.5 Ma

4–10 Ma

△ 539.4 ± 1 Ma
543.3 ± 1 Ma

● 543.9 ± 0.2 Ma

Fig. 13. Duration of Early Cambrian acritarch and faunal zones estimated from the revised time scale for Early Cambrian. The age of the Precambrian–Cambrian boundary is according to Bowring *et al.* (1993) for Siberia (black dot) and Grotzinger *et al.* (1995) for Namibia (triangle). The maximum age of the Middle Tommotian *Dokidocyathus regularis* Zone in Siberia (asterisk) is according to Bowring *et al.* (1993), re-evaluated biochronologically by Vidal *et al.* (1995). Accordingly, the duration of the sub-Tommotian Manykaian Stage (Isachsen *et al.* 1994) or its analogous Nemakit–Daldynian Stage (Grotzinger *et al.* 1995) is also re-evaluated and reduced to only a few million years. Ns = *Nochoroicyathus sunnaginicus* Zone, Dr = *Dokidocyathus regularis* Zone, Dl = *Dokidocyathus lenaicus* Zone in the Tommotian Stage in Siberia. Arrows mark the occurrence of acritarchs in Baltica and Siberia, age-diagnostic for the *Heliosphaeridium–Skiagia* Assemblage Zone. Correlation of acritarch and faunal zones according to Moczydłowska (1991) and Vidal *et al.* 1995..

isotopically younger than Middle Tommotian strata in Siberia and, if the date is accepted, would be coeval to the trilobite-bearing zone(s) in Baltica (*Holmia kjerulfi* Zone) and other parts of Avalonia (*Callavia* Zone). Hence, the age of 530.7±0.9 Ma can not predate the Tommotian–Manykaian boundary as proposed by Isachsen *et al.* (1994). Accordingly, the estimation of duration of the Manykaian Stage as approximately 14 Ma (Isachsen *et al.* 1994) is unlikely, and it derives from the above calculations that it may have lasted no more than a few million years. The same is true of the attribution of a numerical age 530 Ma to the beginning of Tommotian by Grotzinger *et al.* (1995, Fig. 4), which results in an overestimated duration of 14 Ma for the Nemakit–Daldynian Stage (synonymous to the Manykaian). In an Early Cambrian chronology compiled by Tucker & McKerrow (1995), 534 Ma is also regarded the age of the base of the Tommotian (following Bowring *et al.* 1993), whereas in fact it is a maximum age for the Middle Tommotian (Vidal *et al.* 1995; Compston *et al.* 1995).

Regardless of which dates are accepted for the Precambrian–Cambrian boundary (between 539.4±1 Ma and 543.3±1 Ma, Grotzinger *et al.* 1995), the minimum, the maximum age or the mean age, the estimated duration of the two lowermost zones in the Early Cambrian (Fig. 13) is no more than 2–5 Ma per zone. This allows quite precise evaluations of the rates of phytoplankton and metazoan origination and radiation, rates of sedimentation and geochemical turnovers.

Acritarch-based biochronology

Stratigraphic ranges of acritarch species

Many acritarch species recovered in the Silesian successions have well-established stratigraphic ranges with respect to trilobite zones. Their ranges are compiled from the first (FAD) and the last appearance datum (LAD). The Cambrian trilobite zonation used here was established for Scandinavia and the East European Platform (Baltica) and was summarized by Bergström & Gee (1985) and Mens *et al.* (1990). It also applies to the Middle and Upper Cambrian of Newfoundland (West Avalonia) and NW Wales (East Avalonia) belonging to the Acado-Baltic trilobite Province (Westergård 1950; Dean 1985; Dean *in* Martin & Dean 1988; Young *et al.* 1994). The revised names of the trilobite zones *Acadoparadoxides oelandicus* (formerly *Eccaparadoxides oelandicus*) and *Acadoparadoxides pinus* (formerly *Eccaparadoxides pinus*) are here used according to Young *et al.* (1994). The extent of the Lower Cambrian zones was recently slightly modified for the East European Platform in Poland (Moczydłowska 1991) and Scandinavia (Ahlberg & Bergström 1993).

The stratigraphic ranges of the Lower Cambrian acritarch species are given according to Volkova *et al.* (1979) and Moczydłowska (1991), with minor modifications. Compilations of worldwide occurrences of Middle and Late Cambrian species were provided by Welsch (1986), DiMilia (1991) and Albani *et al.* (1991). New references are added, together with slight modifications that result from re-evaluations of their stratigraphic ranges or taxonomic re-assessment (see below). Stratigraphic ranges are summarized in Fig. 14. Only species having the FAD and/or LAD documented by the co-occurrence of macrofossils are here used for further estimation of their relative age.

Numerous Middle–Upper Cambrian species recorded in the Upper Silesia successions are cosmopolitan and have long ranges embracing several trilobite zones (e.g., *Eliasum llaniscum*, *Timofeevia lancarae*, *T. phosphoritica*, *Cristallinium cambriense*, *C. ovillense* and *Multiplicisphaeridium martae*). Their stratigraphic significance lies in the fact that their FAD marks the base of the Middle Cambrian, thus excluding an early Cambrian age. This feature is particularly useful, since many species that occur commonly in transitional Lower–Middle Cambrian strata cross the boundary. Short-ranging species, potentially the most useful for biostratigraphy, are few in the present records. These are *Heliosphaeridium notatum*, *Lophosphaeridium variabile*, *L. latviense* n.comb., and *L. bacilliferum*. *Heliosphaeridium notatum* is known from the Lower Cambrian *Protolenus* Zone or time-equivalent strata in Poland, Latvia, Sweden, Spain and Greenland, but only as rare occurrences (Volkova 1969a, b; Volkova *et al.*1979; Hagenfeldt 1989a; Moczydłowska 1991; Palacios & Vidal 1992; Vidal & Peel 1993). It occurs more frequently in the Middle Cambrian *Acadoparadoxides oelandicus* Superzone in many localities in Poland, Latvia, Lithuania, Russia, Sweden, Finland, Spain and Morocco (see under 'Systematic palaeontology'). This pattern of stratigraphic distribution and abundance is not the result of a biased sampling of Lower and Middle Cambrian rocks. It seems that the *A. oelandicus* Superzone is the 'acme zone' for *H. notatum*. The species is regarded as diagnostic for the Lower–Middle Cambrian transitional interval because of its narrow and well-constrained stratigraphic range and its cosmopolitan palaeogeographic distribution.

Lophosphaeridium latviense n.comb. and *L. variabile* have hitherto only been recorded from the *A. oelandicus* Superzone in Latvia (Volkova 1974; Volkova *et al.* 1979), in Sweden (Hagenfeldt 1989a, b; but see comments under 'Palaeontological descriptions'), and Upper Silesia (this study). *L. bacilliferum* occurs in the uppermost Lower Cambrian and lower Middle Cambrian in Belgium and Sweden (Vanguestaine 1974, 1978; Eklund 1990). These species are very rare and this feature might limit their stratigraphic usefulness. However, a new occurrence of *L. latviense* n.comb. in Spain extends its palaeogeographic distribution and confirms the consistency of its FAD with the base of the Middle Cambrian. The species is recorded in the La Albuera section near Zafra (in the Ossa–Morena Zone according to Lotze 1945) in strata of the lowermost Middle Cambrian (Palacios 1993). The age of the strata (the informal Playon beds) is defined by the trilobites *Parasolenopleura* sp., *Jincella*? sp., and *Solenopleuropsis* cf. *verdiagana*, occurring in the continuously overlying levels correlative to *Paradoxides paradoxissimus* and *P. forchhammeri* Superzones (Liñán *et al.* 1993a, b, 1995).

Particular stratigraphic significance was attributed to *Adara alea*, a species whose short stratigraphic range, as previously recognized (Martin & Dean 1988), was limited to part of the *Paradoxides paradoxissimus* Superzone. The present record provides new evidence suggesting that its range may extend into the Upper Cambrian (see below).

A substantial number of Middle and Upper Cambrian species have stratigraphic ranges that are imprecisely determined within different parts of these series. This refers particularly to species recognized in the Oville Formation in northern Spain. The Oville Formation yielded numerous, taxonomically diverse acritarchs (Cramer &

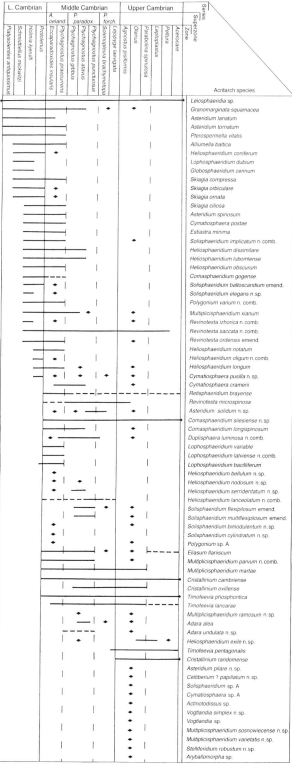

Fig. 14. Stratigraphic succession of acritarch species in the Cambrian of the Upper Silesia area. Cambrian faunal zonation according to Mens *et al.* (1990), Moczydłowska (1991), and Young *et al.* (1994). Stratigraphic ranges of acritarchs revised from various sources. Solid bar marks ranges recognized in respect to trilobite zone, and interrupted bar the ranges inferred micropalaeontologically or through lithostratigraphic correlation. Rhomb indicates ranges inferred in Upper Silesia. Arrow marks extension of range from older or into younger zones. Question mark indicates uncertain extension of range.

Diez Cramer 1972; Fombella 1977, 1978, 1979, 1982, 1986, 1987; Fombella *et al.* 1993) including new species that appear to have had a cosmopolitan palaeogeographic distribution. However, their relative occurrences and ranges within the Oville Formation are poorly documented. The Oville Formation is a succession of shales and glauconitic sandstones of variable thickness (80–800 m), extending through the Cantabrian Mountains in northern Spain (Zamarreño 1972; Aramburu *et al.* 1992). The formation is considered diachronous, spanning Middle Cambrian to presumably Early Tremadoc (Sdzuy 1971b; Aramburu *et al.* 1992). However, only the age of its lower boundary is well constrained by means of trilobites. In a westerly direction, its age gradually becomes younger within the range of the two lower stages of the Middle Cambrian that are equivalent to the *A. oelandicus* and *P. paradoxissimus* Superzones (Sdzuy 1968, 1971b, 1972; Aramburu *et al.* 1992; Liñán *et al.* 1993b). The age of the upper boundary is poorly defined, since there are no diagnostic macrofossils, and it is largely based on interpretation of the acritarch data. The assumption that the upper boundary has a variable age over a broad interval from the Upper Cambrian to Tremadoc was speculative. Acritarchs from the Oville Formation were interpreted by Fombella (1978, 1979, 1982, 1986, 1987; Fombella *et al.* 1993) as indicating different ages for discrete portions of the succession at various localities, thus being strongly diachronous and spanning a Middle Cambrian to Early Tremadoc age. This biostratigraphic assessment was questioned by Vanguestaine & Van Looy (1983), who considered the assemblages described by Fombella (1978, 1979, 1982) as typically Middle Cambrian. They argued that certain species had been misidentified and, as a consequence, that the age of the Oville Formation had been incorrectly inferred. Their remarks were not accepted by Fombella (1986), who maintained that the formation spanned the Upper Cambrian and Tremadoc as well as Middle Cambrian.

The acritarch record in the Oville Formation is problematic in that the sections yielding microfossils were not measured and consequently the exact stratigraphic position of the species within the formation is not known. Very limited information was provided (Fombella 1986, 1987), but from this it is possible to deduce that the sampled intervals occur in the successions referred to the Middle Cambrian *Solenopleuropsis* Zone (Sdzuy 1968, 1971b; Zamarreño 1972). This zone corresponds to the upper part of the *P. paradoxissimus* Superzone (Sdzuy 1971b; Zamarreño 1972; Liñán *et al.* 1993b). Moreover, all the studied sections are located in the central and eastern region of the Cantabrian Mountains (Fombella 1979, 1982), where the Oville Formation has a moderate thickness (106–190 m) and is restricted to the Middle Cambrian (Aramburu *et al.* 1992, Fig. 3). In addition, acritarchs from the Oville Formation were in some cases

incorrectly identified. These include specimens illustrated by Fombella (1982, 1986, 1987; Fombella *et al.* 1993) and attributed to species that are diagnostic for a Late Cambrian and Tremadocian age. None of the illustrated specimens display features diagnostic of taxa to which they were assigned. Some are poorly preserved and unidentifiable, whereas others appear to represent degraded sphaeromorphs. Additionally, leiosphaerids with compressional folds were probably misidentified as presumed Tremadocian spores. One alleged chitinozoan is most likely a fragment of a large sphaeromorph. Taxonomic revision of these acritarchs is not the main purpose of this paper, but they seem to belong to species having long ranges or being characteristic for the Middle Cambrian. In my opinion, there is no evidence of any species indicative of a Late Cambrian or Tremadocian age in the Oville Formation. The biostratigraphic re-evaluation of the taxa reported by Fombella is hampered by the lack of the information concerning the provenance of the specimens (locality and sample) in the plate captions. Based on these arguments, I regard the acritarchs recorded from the Oville Formation to be exclusively Middle Cambrian.

Because of these problems, the upper range of *Comasphaeridium gogense*, recorded as the synonymous *nomen nudum C. filiforme* in the Oville Formation and regarded as Upper Cambrian (Fombella 1979, 1987), is uncertain. The same is true of the LAD of *Cristallinium ovillense, T. lancarae, M. martae, E. llaniscum* and *Heliosphaeridium lanceolatum*, recorded in the Oville Formation and referred to the Lower Tremadoc (Fombella 1978, 1979, 1986).

Species with well-constrained stratigraphic ranges are used to infer the relative ages of assemblages and strata in the Upper Silesian successions. The ranges of new species are based on these inferred ages, and the ranges of certain other species are extended in consequence of the inferred ages. Thus the LADs of *Heliosphaeridium coniferum, H. oligum* and *Solisphaeridium baltoscandium* emend. are in the Middle Cambrian *Acadoparadoxides oelandicus* Superzone (Fig. 14). These species were previously recorded only in the Lower Cambrian (Volkova *et al.* 1979; Moczydłowska 1991; Eklund 1990).

Multiplicisphaeridium xianum, Revinotesta ordensis and *Solisphaeridium implicatum* n.comb. were previously recorded from the Lower and Middle Cambrian (Fombella 1977, 1979; Downie 1982; Moczydłowska 1991). Here I propose that their stratigraphic ranges are extended into an undetermined lower part of the Upper Cambrian (Fig. 14). *Revinotesta izhorica* n.comb. was originally described from the Lower Cambrian but seems to have a longer range embracing the Middle and Upper Cambrian. The ranges of *Comasphaeridium longispinosum, Cymatiosphaera cramerii, Multiplicisphaeridium parvum* n.comb., *Solisphaeridium flexipilosum* emend., and *S. multiflexipilosum* emend., known previously from the

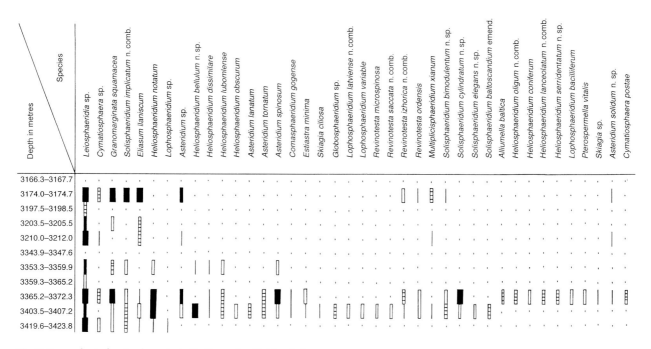

Fig. 15. Range chart of acritarch species in the Sosnowiec IG-1 borehole.

Middle Cambrian, are also extended into the Upper Cambrian (Fig. 14).

Interestingly, the stratigraphic ranges of species of the picoplanktic *Revinotesta* are comparable to those of the morphologically conspicuous and larger species of *Timofeevia*, *Cristallinium* and *Multiplicisphaeridium*. The rarity of records of minute specimens of *Revinotesta* may be due to difficulties in observation and/or preparation methods, but probably not to endemism or scarcity.

The stratigraphic ranges of new species recognized in the Sosnowiec succession can only be estimated as being broadly in the Middle Cambrian *Acadoparadoxides oelandicus* and/or *P. paradoxissimus* Superzone, and/or generally in the Upper Cambrian. In some cases the established ranges are supported by the record of synonymous species elsewhere. The inferred range of *Heliosphaeridium serridentatum* n.sp. in the Sosnowiec succession is the *A. oelandicus* Superzone, whereas in Finnmark (Norway) it occurs in the *P. paradoxissimus* Superzone (Welsch 1986; see under 'Taxonomy'). *Duplisphaera luminosa* n.comb., *Adara undulata* n.sp., and *Comasphaeridium silesiense* n.sp., formally recognized herein, were previously described from the Oville Formation (Fombella 1978, 1979, 1986, 1987; see under 'Taxonomy'). However, their ranges seem to be longer than previously estimated from the Oville Formation (Fig. 14). The revised record of the FAD and LAD of some species and more detailed re-evaluation of their ranges is provided below.

Re-evaluation of stratigraphic ranges of some species

Adara alea and *Cristallinium randomense* occur together in the Sosnowiec succession (at the depth of 3174.0–3174.7 m) but were previously recorded from stratigraphically separated units. Acritarchs recovered in Sosnowiec are regarded as deposited '*in situ*', and there is no evidence of redeposition within the investigated interval of the Sosnowiec succession (see under 'Taphonomy'). The present record provides new evidence indicating that the stratigraphic ranges of some acritarch species are more comprehensive than previously recorded.

The range of *A. alea* was defined in eastern Newfoundland as Middle Cambrian, bracketing the stratigraphic interval comprising the uppermost *Tomagnostus fissus* and *Ptychagnostus atavus* Zone up to the *Ptychagnostus punctuosus* Zone within the *Paradoxides paradoxissimus* Superzone (Martin & Dean 1981, 1984, 1988). The *Adara alea* Range Zone was established by Martin (*in* Martin & Dean 1988) on account of this narrow stratigraphic range only recognized in Newfoundland. Outside the type area, the eponymous species was recorded in Turkey (under the name *Celtiberium geminum*; Erkmen & Bozdoğan 1981; see under 'Taxonomy'), Sweden (as *Adara denticulata* Tongiorgi *in* Bagnoli *et al.* 1988), in Tunisia (Albani *et al.* 1991), and in Spain (Ossa–Morena Zone, Zafra area;

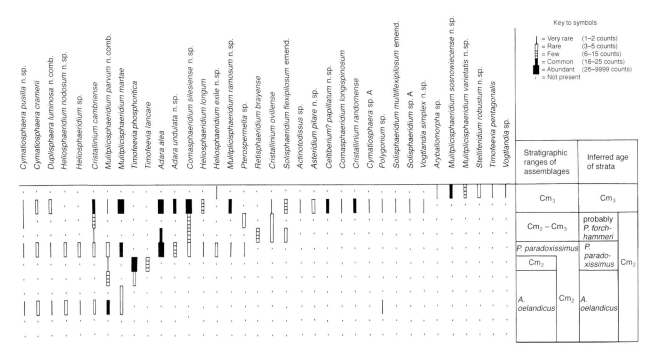

Fig. 15, continued.

T. Palacios, personal communication, 1994, and my own observation). Outside Newfoundland, the occurrence of *A. alea* is stratigraphically controlled only in Sweden, where it occurs in strata referred to the *P. paradoxissimus* Superzone but without more precision (Bagnoli *et al.* 1988), and in Spain, where it co-occurs with the trilobites *Solenopleuropsis*? *verdiagana* and *Jincella*? sp., having ranges also within this superzone (Liñán *et al.* 1993, Liñán *et al.* 1996). In other areas the relative age of *A. alea* was inferred from micropalaeontological data.

Cristallinium randomense occurs throughout the entire Upper Cambrian (*Agnostus pisiformis* Zone to *Acerocare* Zone) extending into the Lower Tremadocian *Clonograptus tenellus* Zone (Martin & Dean 1981, 1988; Welsch 1986). In eastern Newfoundland, the type area of the species, it was recovered from the upper part of the Upper Cambrian, within the interval defined by the *Parabolina spinulosa* Zone and the *Acerocare* Zone (Martin & Dean 1981, 1988; Martin 1982), and the Lower Tremadocian (Martin & Dean 1981). In Norway, *C. randomense* has wider recorded stratigraphic range, appearing in the *Agnostus pisiformis* Zone (Welsch 1986). *C. randomense* has a wide palaeogeographic distribution including, in addition to the above-mentioned areas, Estonia, Belgium, Ireland and (herein) Poland (Volkova 1983, 1990; Meilliez & Vanguestaine 1983; Paalits 1992; Moczydłowska & Crimes 1995; and this paper), and Spain (the Iberian Chains; T. Palacios, personal communication, 1994).

Contrary to previous records, the co-occurrence of *A. alea* and *C. randomense* in the Sosnowiec succession indicates that their ranges overlap and that previously established ranges were incomplete. The record of *A. alea* in eastern Newfoundland is limited to a small area including two nearby localities. At Random Island and Manuels River the species occurs within 1 m and 4 m thick intervals, respectively, of the Manuels River Formation (Martin & Dean 1981, 1984, 1988). *Adara alea* has a widespread palaeogeographic distribution, including West and East Avalonia, Armorica and African Gondwana, and Baltica. Previous records outside Newfoundland have not provided conclusive information on its stratigraphic range. However, the record from one area appears insufficient for establishing the entire stratigraphic range of planktic cosmopolitan species, in particular in such restricted interval. In Upper Silesia, the succession yielding *A. alea* is 36 m thick (Fig. 15). Both in eastern Newfoundland and in Upper Silesia the successions concerned comprise fine-grained, thin-bedded siliciclastics that are not condensed and probably represent similar rates of sedimentation. Because of the greater thickness, it appears obvious that the part of the succession in Upper Silesia containing *A. alea* encompasses a wider stratigraphic interval. This alone, despite the absence of independent stratigraphic control in Upper Silesia, indicates that *A. alea* has a wider range than previously recorded in eastern Newfoundland. Similarly, the range of *C. randomense* was also only partially recognized in eastern Newfoundland, within the same successions as *A. alea*, thus in contrast with its more extensive stratigraphic record in Finnmark, Norway (Welsch 1986). *C. randomense* has a more reliably recognized range, recorded by numerous occurrences independently confirmed as Upper Cambrian in different areas. Therefore the evidence provided by the co-occurrence of *A. alea* and *C. randomense* at the depth of 3174.0–3174.7 m in the Sosnowiec succession is interpreted to suggest that *A. alea* ranges into the lower part of the Upper Cambrian. Alternatively, both species might extend into the uppermost Middle Cambrian. However, additional acritarch genera in this assemblage (*Vogtlandia* and *Actinotodissus*) have their first appearances in Upper Cambrian strata. Hence, an extension of the range of *A. alea* into the Upper Cambrian is favoured here. As a consequence, the *Adara alea* Range Zone (Martin & Dean 1988) should be emended, since the range of the index species was not fully recognized when the zone was established. It seems that the FAD of *A. alea* is in the *Tomagnostus fissus* and *Ptychagnostus atavus* Zone, as documented in eastern Newfoundland, but the species has a definitely wider range recorded in Upper Silesia (Figs. 14 and 15).

Eliasum llaniscum occurs throughout the Middle Cambrian (Welsch 1986; Martin & Dean 1988) and extends into an unspecified portion of the Upper Cambrian, although this part of the range was inferred from micropalaeontological data alone (Moczydłowska & Crimes 1995). Trilobite taxa indicate that in Newfoundland the lower range of *E. llaniscum* is within the Middle Cambrian *Paradoxides bennettii* Zone (Martin & Dean 1983, 1984). This zone corresponds to the upper part of the *Paradoxides oelandicus* Superzone (further referred to as to *Acadoparadoxides oelandicus*; Bergström & Gee 1985, Young *et al.* 1994) and the lowermost part of the *P. paradoxissimus* Superzone in Scandinavia (Martin & Dean 1988). In the East European Platform in Poland, the species was recorded in the lowermost *A. oelandicus* Superzone, the *Eccaparadoxides insularis* Zone, as indicated by the occurrence of the index trilobite species in the Lopiennik IG-1 borehole at the depth of 4715.0–4724.0 m (Moczydłowska 1991 and unpublished data). The first appearance of the species in the Lower Cambrian (Volkova *et al.* 1979; Hagenfeldt 1989a) is poorly documented because the age of the strata was inferred only from lithostratigraphic correlations.

The occurrence of *E. llaniscum* in the uppermost Lower Cambrian, reported as the conspecific *Cymatiosphaera* sp. 1 from Latvia (Ovishi-94 borehole) and referred to the Rausve 'horizon' (Volkova *et al.* 1979), is not supported by any additional data, such as the geologic succession (depth in the borehole) or the macrofossil record within the succession. The basis for the biostratigraphic attribution of the stratigraphic interval in question is uncertain and, by comparison with strata formerly referred to the

Rausve 'horizon' in Lithuania and later referred to the Middle Cambrian (Jankauskas 1980), it may also be questionable. The occurrence of the species in the Middle Cambrian *Acadoparadoxides oelandicus* Superzone in Latvia (Liepaya borehole at the depth of 1363.6 m; Volkova *et al.* 1979), was later confirmed by an additional record in the Vergale-49 borehole at the depth of 1238.0 m (Volkova 1990).

Eliasum llaniscum was also reported from the uppermost Lower Cambrian (*Proampyx linnarssoni* Zone) beds in Sweden underlying the Gotska Sandön Island (Hagenfeldt 1989a). In this section *E. llaniscum* was found in the uppermost part of the File Haidar Formation, just below the boundary with the overlying Middle Cambrian 'oelandicus beds' (Hagenfeldt 1989a, Fig. 3). The lithostratigraphic and biostratigraphic subdivision of the transitional Lower–Middle Cambrian strata in this succession is ambiguous. There are no macrofossils within the interval, whereas ellipsocephalid trilobites occur higher in the 'oelandicus beds'. Lithologically the interval of the uppermost 3 m of strata referred to the File Haidar Formation (and Lower Cambrian) resembles much more the overlying 'oelandicus beds', and there is no obvious reason, either lithostratigraphical or biostratigraphical, to exclude this portion of the succession from the 'oelandicus beds'. The upper File Haidar Formation in Västergötland and Östergötland is regarded as Middle Cambrian in age (*A. oelandicus* Superzone), as inferred from micropalaeontological data (Eklund 1990; Moczydłowska 1991).

The recent record of *Eliasum llaniscum* co-occurring with protolenid–eodiscid trilobites in the inferred uppermost Lower Cambrian in northwestern Wales (Martin *in* Young *et al.* 1994) provides new evidence that the FAD of the species might be in the Lower Cambrian. In Wales, the species occurs in the Hell's Mouth Formation with *Hamatolenus* (*Myopsolenus*) *douglasi*, *Kerberodiscus succinctus* and '*Serodiscus ctenoa*? Rushton' (Young *et al.* 1994). The genus *Hamatolenus* ranges throughout the Lower and lowest Middle Cambrian, whereas elsewhere the two other species are recorded only in the uppermost Lower Cambrian. This was taken to indicate a late Early Cambrian age (Protolenid–Strenuellid Zone) for the Hell's Mouth Formation, though none of the species is sufficiently age-diagnostic to define the Lower–Middle Cambrian boundary (Young *et al.* 1994). Neither does the microfossil assemblage from the Hell's Mouth Formation provide unequivocal evidence for its relative age. Two species in the assemblage, *Skiagia scottica* and *Peramorpha manuelsensis* (Martin in Young *et al.* 1994), are formerly known exclusively from the Lower Cambrian. The illustrated specimen assigned to *S. scottica* (Young *et al.* 1994, Fig. 10m) is not convincingly identified (cf. Downie 1982, Figs. 8k–l, 9a–f; Moczydłowska 1991, Pl. 6E–F; Vidal & Peel 1993, Fig. 14a–d). It is probably more correctly identified with *Skiagia ciliosa*, a species extending through the

Lower Cambrian to the Middle Cambrian *Acadoparadoxides oelandicus* Superzone (Volkova *et al.* 1979; Downie 1982; Moczydłowska 1991). The stratigraphic range of *P. manuelsensis* is poorly constrained, being recorded only in one area (in the Brigus Formation in Newfoundland; Martin & Dean 1983), and thus its entire range might not have been recognized (see above). The species *E. llaniscum* and *Cymatiosphaera ovillensis* (here = *Cristallinium cambriense*) were thus far recorded from Middle Cambrian strata (see under 'Systematic palaeontology'), whereas other species in the assemblage from Wales, *Skiagia insigne*, *Multiplicisphaeridium dendroideum* (= *M. xianum*), *Annulum squamaceum* (= *Granomarginata squamacea*), *Retisphaeridium dichamerum*, *R. howellii*, *Cymatiosphaera capsulara* and *Dichotisphaera gregalis*, have ranges spanning the Lower and Middle Cambrian. *Cymatiosphaera capsulara* Jankauskas 1976 (*in* Jankauskas & Posti 1976) was recorded previously in the Lower Cambrian (Vergale and Rausve 'horizons') in Lithuania and Latvia (Volkova *et al.* 1979), but its co-occurrence with agnostid trilobites in the Ceiriad Formation in Wales documents the extension of the range into the lower Middle Cambrian (Martin *in* Young *et al.* 1994). Additionally, *Dichotisphaera gregalis* is here considered as a non-diagnostic spheromorph (leiosphaerid) lacking any biostratigraphic significance. As a whole, the microfossil assemblage from the Hell's Mouth Formation does not provide evidence for an Early Cambrian age and taxonomically it resembles a Middle Cambrian association of acritarchs.

The extension of the stratigraphic range of *E. llaniscum* into the Upper Cambrian (Fig. 14) is inferred only from micropalaeontological records, since the species occurs in assemblages indicative of a Late Cambrian age (Moczydłowska & Crimes 1995, and present record). Additionally, the species occurs with *C. randomense* in an assemblage of acritarchs from the Umbria Pipeta Formation in Sierra Morena, Spain (Mette 1989). Although the assemblage was considered to be Middle Cambrian by Mette (1989), the occurrence of *C. randomense*, which has a well-established range in respect to trilobite zones and has only been found in Upper Cambrian strata, suggests rather a Late Cambrian age. The Umbria Pipeta Formation was previously referred to the Upper Cambrian (Schneider cited in Mette 1989).

The occurrence of *Cristallinium cambriense* in Lower Cambrian strata elsewhere is also uncertain. A few poorly preserved specimens were found in the upper File Haidar Formation (Lingulid Sandstone member) in Västergötland (Moczydłowska & Vidal 1986), and one specimen was recorded in Östergötland (Eklund 1990). The Lingulid Sandstone member embraces transitional Lower–Middle Cambrian strata (Eklund 1990; Moczydłowska 1991) and is overlain by a glauconitic sandstone that is succeeded by the '*oelandicus* mudstone' (Eklund 1990).

Whether the Lower–Middle Cambrian boundary should be placed within the Lingulid Sandstone member is a matter of controversy, since there are no age-diagnostic macrofossils in this condensed succession (Eklund 1990 and references therein). Based on acritarch evidence Eklund (1990) regarded the topmost part of the Lingulid Sandstone in Östergötland as Middle Cambrian in age. Hence, thus far, the only specimen of *C. cambriense* recorded by him occurs below the proposed boundary (Eklund 1990). This occurrence is not taken as definitive evidence for locating the FAD of the species in the Lower Cambrian because of the biostratigraphic difficulties in recognizing the exact position of the Lower–Middle Cambrian boundary in the succession.

Species regarded here as synonymous with *Cristallinium cambriense*, described from Lithuania by Jankauskas (1976) under the generic name *Cymatiosphaera* (see under 'Taxonomy'), were regarded as early Cambrian in age (Jankauskas 1976; Volkova *et al.* 1979). The relative age of the strata concerned was based entirely on the occurrence of microfossils, without any additional form of age control. The Lakaj Formation was considered to be Early Cambrian and to correlate with the Vergale–Rausve 'horizons' (Jankauskas 1974, 1976; Volkova *et al.* 1979), though numerous recorded acritarch species have stratigraphic ranges spanning the upper Lower Cambrian – lower Middle Cambrian. Additional species from this formation were believed to be new taxa (*Cymatiosphaera favosa* Jankauskas 1976 and *Multiplicisphaeridium vilnense* (Jankauskas 1976) Jankauskas 1979), but were later recognized by Jankauskas (1980) as junior synonyms of the Middle Cambrian species *Cymatiosphaera ovillensis* Cramer & Diez 1972 (= *Cristallinium cambriense*) and *Multiplicisphaeridium lancarae* Cramer & Diez 1972 (= *Timofeevia lancarae*), respectively. Subsequently, Jankauskas (1980, pp. 132, 133) re-evaluated the relative age of the Lakaj Formation as Middle Cambrian.

Occurrences of *Heliosphaeridium lanceolatum* in various palaeo-regions (e.g., Belgium, Turkey, Russia and Poland) are consistent with a Middle Cambrian age (Vanguestaine 1974, 1978, 1992; Erkmen & Bozdoğan 1981; Volkova 1990; Fig. 14). The species was also reported from the Oville Formation at Adrados in Spain (Fombella 1986). An Early Tremadoc age was inferred for this occurrence on the basis of a microfossil assemblage consisting of acritarchs and 'chitinozoans' alone. However, no information is available concerning the stratigraphic level of the microfossils within this succession, which exceeds 70 m in thickness. Consequently, only the Middle Cambrian stratigraphic range of *H. lanceolatum* is confidently established, whereas its extension into the Lower Tremadocian is extremely uncertain (see under 'Stratigraphic ranges').

Biochronological constraints on the Sosnowiec succession

The relative age of the acritarch assemblages from each sample was considered separately. The acritarch assemblages form a younging-upwards succession. The lowermost strata, below 3423.8 m, may be regarded as possible transitional Lower to Middle Cambrian (Fig. 6). The bulk of the succession is Middle Cambrian, and only its upper part is referred to the Upper Cambrian.

The acritarchs recovered from 3419.6–3423.8 m consist of a few commonly known species, each represented by rare specimens (Fig. 15). Stratigraphically significant species are *Heliosphaeridium notatum* and *Eliasum llaniscum*. The interval bracketed by the overlapping ranges of the two species is in the lower Middle Cambrian, equivalent to the *Acadoparadoxides oelandicus* Superzone. This is the relative age inferred for the acritarch assemblage and the sampled interval in the Sosnowiec IG-1 borehole.

The acritarchs from 3403.5–3407.2 m are taxonomically diverse and include a number of species with ranges that are well-established with reference to trilobite zones. Most of the species occur in the upper Lower Cambrian *Holmia kjerulfi* and *Protolenus* zones and the lower Middle Cambrian *A. oelandicus* Superzone. These include various species of *Asteridium* and *Heliosphaeridium*, *Revinotesta ordensis*, *Estiastra minima*, *Comasphaeridium gogense* and *Skiagia ciliosa*. Their co-occurrence with *Eliasum llaniscum* and *Revinotesta microspinosa*, having their FAD in the Middle Cambrian *A.oelandicus* Superzone, suggests such an age for the entire acritarch assemblage. This is also supported by the presence of *Lophosphaeridium variabile* and *L. latviense* n.comb. in the assemblage, species potentially age-diagnostic for the *A. oelandicus* Superzone (see under 'Stratigraphic ranges'). *Asteridium lanatum*, present only in this assemblage in the Sosnowiec succession, has its LAD in the *Eccaparadoxides insularis* Zone of the *A. oelandicus* Superzone (Hagenfeldt 1989a; Fig. 14). It may be taken to suggest the exact stratigraphic position of the sampled interval. *Heliosphaeridium notatum* is a dominant species in the assemblage, reaching a frequency of over 33% of the total number of specimens among all the 24 species present. This represents its highest abundance in relation to other stratigraphic levels in the Sosnowiec succession (Fig. 15). Such a pattern of maximum abundance in the distribution of the species coincides with the inferred early Middle Cambrian age representing the 'acme zone' of *H. notatum* (see under 'Stratigraphic ranges').

Other species in the assemblage (*Granomarginata squamacea*, *Solisphaeridium implicatum* n.comb., *Revinotesta saccata* and *Multiplicisphaeridium xianum*) have ranges that embrace the Lower, Middle and Upper Cambrian. The ranges of *Solisphaeridium baltoscandium* emend., and

Revinotesta izhorica n.comb., previously recorded only from the Lower Cambrian are here inferred to extend into the Middle Cambrian (the former species) and Upper Cambrian (the latter species, Fig. 14). The new species *Solisphaeridium bimodulentum* n.sp. and *S. cylindratum* n.sp., recognized in the microfossil assemblage from the interval of 3403.5–3407.2 m, are inferred to have a first recorded appearance in the *A. oelandicus* Superzone, *E. insularis* Zone.

The acritarch assemblage from 3365.2–3372.3 m resembles the assemblage from underlying strata (3403.5–3407.2 m). The majority of species have ranges embracing the upper Lower Cambrian with last occurrences in the Middle Cambrian (*A. oelandicus* Superzone). These include *Heliosphaeridium notatum*, *Asteridium spinosum*, *A. tornatum*, *Alliumella baltica*, *Estiastra minima*, *Cymatiosphaera postae*, *Pterospermella vitalis* and *Lophosphaeridium bacilliferum*. *H. notatum* is the most abundant species within this group. A second group of species includes *Eliasum llaniscum*, *Multiplicisphaeridium martae*, *M. parvum*, *Cymatiosphaera cramerii* and *H. lanceolatum*. These are stratigraphically significant, having their lowest documented occurrences in the lower Middle Cambrian. The FAD of *E. llaniscum* and *M. parvum* is in the *A. oelandicus* Superzone (Hagenfeldt 1989b; Fig. 14). *M. parvum* is known exclusively from this zone in Sweden and Finland (Hagenfeldt 1989b), but its range seems not to have been fully determined there. The present record widens the palaeogeographic distribution of the species outside Scandinavia and extends its range up to a lower portion of the Upper Cambrian (Fig. 14). The first appearance of *C. cramerii* was recognized in the lower Middle Cambrian *Ellipsocephalus hoffi* Zone by Slaviková (1968) but in the stratigraphically upper *Eccaparadoxides pusillus* Zone by Vavrdová (1976) in the Bohemian Massif. The Middle Cambrian strata in the Bohemian Massif have been correlated with the *Paradoxides paradoxissimus* Superzone (Öpik 1979). In Newfoundland the first appearance of the species is in strata of the *Paradoxides bennettii* Zone (Martin & Dean 1983, 1984), corresponding to the upper part of the *Acadoparadoxides oelandicus* and lower *Paradoxides paradoxissimus* Superzones (Martin & Dean 1984, 1988). The base of the stratigraphic range of *M. martae* is less precisely established. It was referred to the lower Middle Cambrian *Eccaparadoxides pusillus* Zone (Vavrdová 1976) corresponding to the *Paradoxides paradoxissimus* Zone (Fatka 1989) in the Bohemian Massif, and the *Paradoxides paradoxissimus* Zone in Norway (Welsch 1986). The range of the species in other worldwide occurrences was attributed more generally to the lower Middle Cambrian or Middle Cambrian (Vanguestaine & Van Looy 1983; Fombella 1986; Volkova 1990; see under 'Taxonomy').

The overlapping interval of the stratigraphic ranges of these two distinctive groups of species is in the Middle

Cambrian *A. oelandicus* Superzone. This is the proposed age for the strata in question in the Sosnowiec succession. The remaining species in the assemblage are new or have ranges spanning the Lower to Upper Cambrian, such as *Granomarginata squamacea* and *Solisphaeridium implicatum* n.comb. Among them the most abundant is *G. squamacea*. *Duplisphaera luminosa* n.comb. also has a long stratigraphic range spanning the Middle Cambrian (in the Oville Formation, Fombella 1978, 1987; but see under 'Stratigraphic ranges') and extending into the Upper Cambrian as inferred in the Sosnowiec formation (Figs. 14 and 15).

All species in the assemblage from 3353.3–3359.3 m, except *Timofeevia phosphoritica*, are also recorded in the assemblages from the underlying strata. Occurrence of *T. phosphoritica* in this stratigraphic level is the lowest in the Sosnowiec succession. The overlapping ranges of all species are in the *A. oelandicus* Superzone.

The acritarch assemblage from 3343.9–3347.6 m consists of only three species, ranging from Middle to Upper Cambrian. This interval is sandwiched by strata of inferred Middle Cambrian age (Fig. 6), and an age within the *P. paradoxissimus* Superzone is suggested.

The majority of species in the assemblage from the cored interval of 3210.0–3212.0 m occur in the Middle and Upper Cambrian, and a few extend into the Lower Tremadoc. One species (*Heliosphaeridium longum*) was previously recorded only from the Lower Cambrian and the lowermost Middle Cambrian. A few species are new, and so their ranges could be only partly recognized from the synonymous taxa. Within the group of species occurring throughout the Middle and Upper Cambrian, particular stratigraphic significance is attributed to *Adara alea*. Its FAD coincides with the *Tomagnostus fissus* and *Ptychagnostus atavus* Zone of the *Paradoxides paradoxissimus* Superzone (Martin & Dean 1988), but it is interpreted to extend into the Upper Cambrian (Fig. 14; see under 'Re-evaluation of stratigraphic ranges'). *A. alea* is one of the dominant species in the present assemblage. It constitutes 38% of the total number of specimens, whereas *M. martae* makes up 15% of all specimens and the other nineteen species in the assemblage are represented by a few per cent each. Among the new species in this assemblage, *Asteridium solidum* n.sp., *Comasphaeridium silesiense* n.sp. and *Adara undulata* n.sp. were formerly described as *nomina nuda* and have a number of previously recorded occurrences. Thus, their stratigraphic ranges are already at least partly known. These species range from Middle to Upper Cambrian (Fig. 14).

The overlapping ranges of acritarchs from 3210.0–3212.0 m embrace the Middle Cambrian *P. paradoxissimus* Superzone to the Upper Cambrian. Strata in the Sosnowiec borehole form a normal stratigraphic succession, and the first record of *A. alea* at this level is interpreted to indicate a Middle Cambrian age, equivalent to the upper *P. paradoxissimus* Superzone, for the entire assemblage and the sampled strata.

Within the assemblage from 3203.5–3205.5 m ,two species, *Cristallinium ovillense* and *Solisphaeridium flexipilosum* emend., have their lowest stratigraphic occurrence in this interval and *Retisphaeridium brayense* the only record in the Sosnowiec formation (Fig. 15). The other species have continuous records in the underlying and overlying strata. The stratigraphic ranges of all species are within the Middle and Upper Cambrian.

The assemblage from 3197.5–3198.5 m consists of *C. cambriense* and *C. ovillense*, commonly co-occurring in the upper Middle Cambrian and lower Upper Cambrian (Figs. 14 and 15), and two new species. More precisely, the relative age of the sampled strata can be assessed only in the context of the whole succession and is interpreted as probably time-equivalent to *P. forchhammeri* Superzone (Figs. 15 and 6).

The microfossil assemblage from 3174.0–3174.7 m is taxonomically diverse, consisting of 34 species with the greatest abundance in the entire succession. Ranges of some species in the assemblage conflict. Based on observations of sedimentary structures and the state of preservation of microfossils, it is concluded that the assemblage accumulated *in situ* (see under 'Taphonomy'). Therefore the apparently conflicting ranges are explained in terms of previously incompletely known stratigraphic records. Accordingly, the stratigraphic ranges of certain species are herein extended (Fig. 15).

A number of species in this assemblage, including *Eliasum llaniscum*, *Cristallinium cambriense*, *C. ovillense* and *Multiplicisphaeridium martae*, that were also recorded in the previous samples, have long stratigraphic ranges extending through the Middle–Upper Cambrian (Fig. 14). *Comasphaeridium silesiense* n.sp. can be added to this group, and although a new species, it has a well-recognized range. This species is very abundant, and together with *Eliasum llaniscum* and *M. martae* it dominates the assemblage. In contrast, *C. randomense* only occurs at this particular depth interval in the Sosnowiec borehole and has a shorter stratigraphic range bracketing Upper Cambrian – lowermost Tremadocian. Because of its well-recognized range and common distribution, *C. randomense* is considered as an age-diagnostic species. The presence of this species suggests that the assemblage cannot be older than Late Cambrian. This is supported by the appearance of *Vogtlandia simplex* n.sp. in the assemblage. The genus *Vogtlandia* was never reported from strata older than Late Cambrian (Volkova 1990), and most of the species are in fact Ordovician in age (Fensome *et al.* 1990).

Along with *A. alea* several other species in the above assemblage (e.g., *Multiplicisphaeridium parvum* n.comb., *C. cramerii*, *S. flexipilosum* emend., *S. multiflexipilosum* emend., *C. longispinosum*) were previously recorded only in the Middle Cambrian, whereas two species (*Multipli-*

cisphaeridium xianum and *Revinotesta ordensis*) were recorded in Lower and Middle Cambrian strata. *M. parvum* n.comb., and *C. longispinosum* are thus far known from a limited palaeogeographic distribution (a few localities in Sweden and Finland) in the *A. oelandicus* Superzone (Hagenfeldt 1989b), whereas *S. flexipilosum* emend., and *S. multiflexipilosum* emend. are known from the Czech Republic (Vavrdová 1976) in strata time-equivalent to the *Paradoxides paradoxissimus* Superzones (Öpik 1979). Because of the sparse records, it is possible that their complete stratigraphic ranges are still not fully recognized. The biostratigraphic evaluation of this assemblage is based on the appearance of *C. randomense*, which has well-established record in the Upper Cambrian, and therefore the range of the former species probably also extends into a lower portion of the Upper Cambrian .

Multiplicisphaeridium xianum and *Revinotesta ordensis* have a cosmopolitan palaeogeographic distribution in the Lower and Middle Cambrian. By comparison with the associated species in the assemblage, it seems that their ranges also extended into the Upper Cambrian. Comparably, the range of *Revinotesta izhorica* was previously limited to the Lower Cambrian (Jankauskas 1975), to strata equivalent to the *Holmia kjerulfi* Zone but in the palaeogeographically restricted distribution in Baltica.

Among the new species in the assemblage, *Adara undulata* n.sp. was previously recorded, as a synonymous species, within the presumably lower Middle Cambrian in the Oville Formation. In Silesia it occurs also at a depth of 3210.0–3212.0 m in the inferred Middle Cambrian strata (*P. paradoxissimus* Superzone). The new species that were also recorded in the underlying intervals of the Sosnowiec succession, *Solisphaeridium bimodulentum* n.sp., *Cymatiosphaera pusilla* n.sp. and *Multiplicisphaeridium ramosum* n.sp., seem to have long stratigraphic ranges (Fig. 14). This is inferred from the known stratigraphic ranges of the other species in the assemblages. The species *Vogtlandia simplex* n.sp., *Asteridium pilare* n.sp., *Celtiberium? papillatum* n.sp., and *Solisphaeridium* sp. A are recorded exclusively in the present interval of strata. The stratigraphic range of *Vogtlandia simplex* n.sp., inferred from the assemblage, is consistent with the record of genus *Vogtlandia* that is thus far known to be restricted to the Late Cambrian – Early Ordovician (Volkova 1990; Fensome *et al.* 1990). In summary, the microfossil record from the interval of 3174.0–3174.7 m in the Sosnowiec borehole is interpreted as indicative of a Late Cambrian age for the concerned strata.

In the assemblage from 3166.3–3167.7 m, *Timofeevia pentagonalis* is stratigraphically significant, ranging through the topmost Middle Cambrian and Upper Cambrian (Fig. 14). Other species are new or assigned only to genera (Fig. 15). However, the record of *Stelliferidium* and *Vogtlandia*, genera which appear in the Upper Cambrian and are more diverse in the Lower Ordovician (Vavrdová

1976, 1982; Welsch 1986; Martin & Dean 1988; Bagnoli *et al.* 1988; Volkova 1990; Fensome *et al.* 1990; DiMilia 1991; Albani *et al.* 1991), is also age-diagnostic. Previously, only one species of *Stelliferidium* (*S. pingiculum*) has been recorded in the Upper Cambrian *Olenus* and *Parabolina spinulosa* zones (Martin & Dean 1988). A single species of *Vogtlandia* (*V. notabilis*) was reported from the Upper Cambrian *Leptoplastus* and *Peltura* zones (Volkova 1990). *Vogtlandia simplex* n.sp. is recorded in the Sosnowiec Formation in the underlying strata inferred to be Upper Cambrian (see above). The genus *Aryballomorpha*, occurring in the present assemblage, was until now known from the Tremadocian (Martin 1984; Dean & Martin 1982; Martin & Yin 1988; Volkova 1993). *Heliosphaeridium exile* n.sp. was recorded in an additional stratigraphic level in the Sosnowiec formation (Fig. 15), and as synonymous species in Ireland (Gardiner & Vanguestaine 1971; see under 'Systematic palaeontology'). In both instances its stratigraphic range might be only indirectly inferred from the ranges of co-occurring microfossils, because of lack of the age control by other means. Its range embraces Middle Cambrian (*P. paradoxissimus* Superzone, in Sosnowiec) and the upper part of the Upper Cambrian (in Ireland; Moczydłowska & Crimes 1995).

Based on the well-recognized range of *T. pentagonalis* and indirect evidence provided by the genera ranging from Upper Cambrian to Ordovician, the relative age of the strata in question is inferred to be late Cambrian.

Relative age of the Goczałkowice succession

Acritarch assemblages recorded in the Goczałkowice IG-1 borehole (Fig. 16) are not particularly rich, but they are stratigraphically significant. In addition to Lower Cambrian trilobites (Orłowski 1975; Fig. 8 herein), the acritarchs allow part of the succession to be dated independently as Middle Cambrian. The lowermost occurrence of acritarchs is in the marine deposits immediately overlying the haematitic beds, at the depth of 3022.9–3025.0 m (Fig. 8). The sample yielded only a few specimens of acritarchs, identified as *Globosphaeridium* sp. and a number of undiagnostic leiosphaerids. The genus *Globosphaeridium* is not known from strata older than the *Skiagia ornata – Fimbriaglomerella membranacea* acritarch Zone, which is the time-equivalent of the *Schmidtiellus mickwitzi* trilobite Zone (Moczydłowska 1991; Alhberg & Bergström 1993). The deposits at this stratigraphic level are therefore probably not older than this zone.

The acritarch assemblage from the depth of 2969.2–2973.7 m consists of four species (*Skiagia ciliosa, S. ornata, Globosphaeridium cerinum* and *Lophosphaeridium*

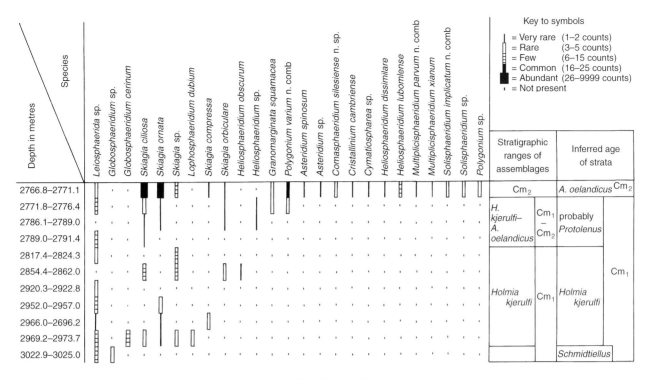

Fig. 16. Range chart of acritarch species by lowest appearance in the Goczałkowice IG-1 borehole.

dubium), with additional microfossils identified to generic level (Fig. 16). The co-occurrence of these species is restricted to the *Heliosphaeridium dissimilare – Skiagia ciliosa* Zone, i.e. the time-equivalent of the *Holmia kjerulfi* Assemblage Zone (Moczydłowska 1991; Ahlberg & Bergström 1993; Fig. 14). The lowermost occurrence of *S. ciliosa* at this stratigraphic level marks the lower boundary of the acritarch zone and, by default, its correlative trilobite zone in the Goczałkowice succession (Fig. 8). Acritarchs from the overlying interval of 2854.4–2969.2 m (Fig. 16) represent common Early Cambrian species such as *S. orbiculare, S. ornata, S. compressa, S. ciliosa* and *Heliosphaeridium obscurum*, with stratigraphic ranges overlapping within the *Heliosphaeridium dissimilare – Skiagia ciliosa* and *Volkovia dentifera – Liepaina plana* Zones. The latter zones are time-equivalent to the *Holmia kjerulfi* and *Protolenus* Zones (Moczydłowska 1991; Ahlberg & Bergström 1993, Fig. 14). Within the succeeding strata containing trilobites (Fig. 8) only poorly preserved acanthomorphic and sphaeromorphic specimens were recorded at the depth of 2817.4–2824.3 m. A few acritarch species recovered from the overlying interval (2771.8–2791.4 m) have longer stratigraphic ranges embracing the *Holmia kjerulfi* to *Acadoparadoxides oelandicus* Zones.

The assemblage from a depth at 2766.8–2771.1 m is taxonomically diverse and consists of numerous acritarch species known previously within the range of the *Holmia kjerulfi* and *A. oelandicus* Zones (*Skiagia compressa, S. cil-*

iosa, Polygonium varium and *Asteridium spinosum*), and some species with more comprehensive ranges (*Multiplicisphaeridium xianum, Heliosphaeridium dissimilare* and *H. lubomlense*). Stratigraphically the most significant taxa in this assemblage are *Cristallinium cambriense, Comasphaeridium silesiense* n.sp. and *Multiplicisphaeridium parvum* n.comb., having their first appearance in the *A. oelandicus* Superzone and extending to the Upper Cambrian (Fig. 14). The overlap of the stratigraphic ranges of all acritarch species in the assemblage is limited to the *A. oelandicus* Superzone, and this is the inferred relative age of the uppermost strata in the Goczałkowice succession. Accordingly, the range of *Skiagia ornata*, present in the assemblage but previously recorded only in the Lower Cambrian (Volkova *et al.* 1979, Moczydłowska 1991), is thus proven to be wider and extended into the *A. oelandicus* Superzone.

The present acritarch record, combined with the trilobite evidence, allows refinement of the biozonation of the Goczałkowice succession (Fig. 8). The diagnostic Early Cambrian trilobites *Strenuaeva primaeva* (Brögger), *Ellipsocephalus nordenskjoeldi* Linnarsson and *Schmidtiellus panowi* (Samsonowicz) occur within the interval of 2793.0–2850.4 m (Biernat *et al.* 1973; Orłowski 1975). Inarticulate brachiopods, mostly lingulids, have been recorded at five different levels between the depths of 2786.1 m and 2948.8 m (Biernat *et al.* 1973). Their biostratigraphic significance is limited since they possess wide stratigraphic ranges, comprising the Lower Cambrian to Ordovician. Unidentified simple burrows, comprising *Bergaueria* sp., occur throughout the succession from a depth of 3032.0 m up to the sub-Devonian unconformity at 2765.0 m (Kotas 1973a; Figs. 8 and 9). The entire sub-Devonian terrigenous succession at Goczałkowice was previously considered to be Lower Cambrian (Kotas 1973a, 1982a, b; Kowalczewski 1990), or partly Precambrian (?Vendian) in its basal portion (Orłowski 1975).

Orłowski (1975) recognized the *Holmia* zone in the part of the succession at 2765.0–2860.0 m, whereas underlying strata (2860.0–3039.0 m) he referred to the sub-*Holmia* zone. The haematitic beds within the interval 3039.0–3129.2 m and beneath the 'fossiliferous' strata were regarded as Precambrian. As argued above, the present find of acritarchs implies that the upper part of the succession referred to as the '*Holmia* zone' (above the occurrence of trilobites) by Orłowski (1975) is time equivalent to the *A. oelandicus* Superzone, and probably *Protolenus* Zone (Fig. 8). The formerly established position of the boundary between the *Holmia* and sub-*Holmia* zones at the depth of 2860.0 m (Orłowski 1975) is not supported by evidence from fossils and it seems to have been chosen by convenience. Orłowski (1975) quoted the occurrence of the *Skolithos*-type burrows in the interval of 2957.0–3039.0 m in beds attributed to the sub-*Holmia* zone. This is, however, inconsistent with the borehole log, in which the occurrence of *Skolithos* was reported from the other portion of the succession, i.e. the interval of 3039.0–3129.2 m, attributed to the informal '*Skolithos* member' (Kotas 1973a, b). This lithostratigraphic unit and its assignment to the Lower Cambrian has been subsequently retained by Kotas (1982a, b) and Kowalczewski (1990), contrary to the Vendian age proposed by Orłowski (1975). The occurrence of *Skolithos* in the Goczałkowice succession was not observed during the course of the present examination of the drillcore. In any case, this fossil lacks major biostratigraphic significance because it has a comprehensive stratigraphic range within the Upper Vendian to post-Palaeozoic (Droser 1991; Aramburu *et al.* 1992; Bottjer & Droser 1994).

A few acritarch species previously reported from the Goczałkowice succession (from the interval of 2766.0–2973.7 m) have overlapping ranges within the Lower–Middle Cambrian (Moczydłowska *in* Kowalczewski *et al.* 1984). The occurrence of microfossils in the metamorphosed portion of the succession (depth of 3177.6–3180.6 m) was erroneously reported by Moczydłowska (in Kowalczewski *et al.* 1984) on a clearly contaminated sample provided for the micropalaeontologic analysis to the author. This is revealed by the present study based on the examination and repeated sampling of the drillcore. The error was unfortunately retained in subsequent studies (Kowalczewski 1990).

Based on the occurrence of age-diagnostic acritarchs and trilobites, part of the investigated succession between the depths of 2793.0 m and 2973.7 m is here referred to the *Heliosphaeridium dissimilare – Skiagia ciliosa* acritarch Zone, being time-equivalent to the *Holmia kjerulfi* Assemblage Zone (Figs. 8 and 16). Its lower boundary is recognized at the level of the lowermost occurrence of *Skiagia ciliosa* in the succession, whose first appearance coincides with the base of this zone (Moczydłowska 1991; Ahlberg & Bergström 1993). The upper boundary is located above the level of the highest recorded occurrence of the trilobite species diagnostic for the zone, i.e. *Schmidtiellus panowi*, which has a concurrent range with *Holmia kjerulfi* (Orłowski 1975).

The boundary between the Lower and Middle Cambrian in the Goczałkowice succession is here placed between the depth of 2793.0 m, where the Early Cambrian trilobite *Schmidtiellus panowi* occurs, and 2771.1 m, above which the first occurrence of Middle Cambrian acritarchs was recorded. The above-mentioned interval has a thickness of 22 m and consists of mudstones and sandstones that were deposited in a shallow shelf environment. No sedimentary break has been observed within this almost entirely cored portion of the succession (Fig. 8), though the presence of a hidden minor hiatus cannot be excluded at a level marked by a shallowing-up depositional trend within quartz sandstones. However, the continuous distribution of the same acritarch species below and above the quartz beds suggests uninterrupted sedimentation. The presence of a diverse age-diagnostic acritarch assemblage at the depth of 2766.8–2771.1 m suggests an important evolutionary bio-event at the beginning of Middle Cambrian. This record is particularly relevant because fossiliferous transitional successions spanning the Lower–Middle Cambrian boundary are rare. It could be inferred that the Goczałkowice succession may embrace the *Protolenus* Zone, although unrecognizable in the lack of the diagnostic fossils.

The lower part of the Goczałkowice succession is here referred to the *Schmidtiellus mickwitzi* Zone. Its recognized boundaries are, however, only approximate because of a sparse fossil record. The shallowing-up episode and an inferred period of non-deposition or starved deposition is marked by phosphoritic conglomerates at the depth of 2973.5–2973.7 m (Fig. 8) and may represent the upper limit of this zone. The lower boundary of the zone is probably located at the level of the lowermost occurrence of bioturbated marine deposits at the depth of 3032.0 m.

A sharp lithological contact exists at the depth of 3032.0 m between bioturbated marine deposits and underlying haematitic deposits accumulated in near-shore, tidally influenced, and on-shore/alluvial environments. No erosional surface is present at this level, although it is possible that the contact may contain a stratigraphic gap. In any case, the Goczałkowice succession represents a transgressive sedimentary sequence, spanning alluvial–costal into lower offshore shelf environments. The succession may contain several hiati, although they are most probably relatively short-lasting. It is very likely that paralic and onshore deposits (at the depth of 3032.0–3110.5 m) accumulated during earliest Cambrian times. Consequently, the underlying conglomerates of the Pszczyna member might be assumed to have accumulated during Late Neoproterozoic times and are here regarded as representing a post-Cadomian molasse.

Tentative estimate of the relative age of the Potrójna succession

The fossil record of the Jaszczurowa formation in the Potrójna IG-1 borehole is generally poor and includes scarce acritarchs and trace fossils (Fig. 17). The trace fossils consist of simple burrows of *Skolithos*-type and *Monocraterion* sp. occurring at several levels within the succession (Ślączka 1976, 1985c). Their biostratigraphic significance is very limited because of the comprehensive stratigraphic ranges (Bottjer & Droser 1994; Jensen 1996). Acritarchs were recovered only at 3356.3–3363.5 m (Moczydłowska *in* Kowalczewski *et al.* 1984; Fig. 17). The ranges of the age-diagnostic species of *Timofeevia* in this association (*T. phosphoritica*, *T. lancarae* and *T. pentagonalis*) overlap within the Upper Cambrian (Fig. 14). An additional taxon, *Multiplicisphaeridium varietatis* n.sp., is recorded also in the Sosnowiec borehole in the portion of the succession inferred to be Upper Cambrian (Fig. 15). Based on the acritarch record, the relative age of the upper part of the Jaszczurowa formation (at a depth of 3356.3–3363.5 m) is inferred to be Late Cambrian.

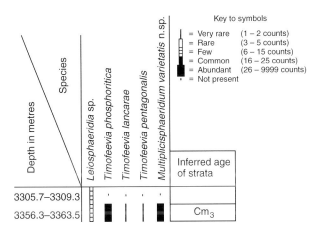

Fig. 17. Range chart of acritarch species by lowest appearance in the Potrójna IG-1 borehole.

Palaeobiogeography

Cambrian acritarchs are known worldwide in various marine shelf depositional settings (Vanguestaine 1991; Moczydłowska 1995b, 1997; see compilation under 'Palaeontological descriptions'), and from areas considered to have been part of different palaeocontinents (Figs. 18 and 19). These include Baltica (encompassing Baltoscandia and the East European Platform), Gondwana and marginal terranes of East Avalonia (southern Ireland, southern Britain, the Brabant Massif and the Ardennes, Lusatia and Upper Silesia), West Avalonia (eastern Newfoundland, Nova Scotia, New England) and Armorica (France, Sardinia, the Bohemian Massif), and Laurentia (Greenland, Svalbard, Scotland, the Mackenzie Mts, Alberta and Tennessee). In Laurentia, Siberia (the Anabar uplift), South China (Yangtze Platform) and Australia (Amadeus Basin), acritarchs have been reported only from the Lower Cambrian. Discrete acritarch species, and in some instances entire assemblages, occur in areas that are now widely separated but which evidently were closer during Cambrian times (Moczydłowska 1995b). The recognition of Cambrian palaeocontinents and associated terranes is based on palaeomagnetic data, climatic sedimentary indicators and faunal distributions following palaeogeographic reconstructions by Scotese & McKerrow (1990), Torsvik *et al.* (1991), Courjault-Radé *et al.* (1992), McKerrow & Cocks (1995) and Dalziel (1995 *in* Palmer & Rowell 1995). The present palaeogeographic reconstruction is slightly modified from the cited sources and complemented by acritarch distributions (Moczydłowska 1995b; Fig. 19 herein). The Upper Silesia terrane is recognized as the distal segment of East Avalonia on the basis of tectonic similarities of Proterozoic basement complexes and the distribution of Cambrian acritarchs and trilobites (Moczydłowska 1995b, 1997; Fig. 18 herein, see under 'Tectonic implications').

Cambrian acritarch associations from Upper Silesia are comparable to assemblages previously reported elsewhere in Baltica, Avalonia, Armorica and Laurentia. However, the strongest taxonomic resemblance is with Baltica and Iberia. The most complete record of Lower, Middle and Upper Cambrian acritarchs is known from Baltica, where they are reported in numerous sites in platform successions and in the Caledonian fold belt. In the latter they are recognized in areas extending for southern Norway to Finnmark. Acritarch biochronology in Baltica is well calibrated in respect to trilobite zonations. Taxonomic affinities of Lower Cambrian acritarch associations from Upper Silesia are evident with those in the Lublin Slope (Poland) and Lake Mjøsa (southern Norway), whereas the recognized Middle and Upper Cambrian succession displays similarities with central Sweden, Finnmark (northern Norway), and the Moscow syneclise. In terms of diversity and taxonomic composition, the acritarch

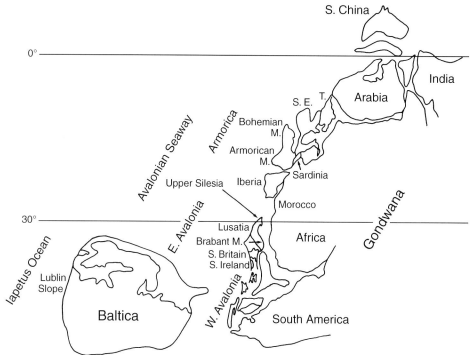

Fig. 18. Early Cambrian palaeobio-geographic reconstruction showing the location of Upper Silesia within Eastern Avalonia terranes at the margin of Gondwana.
S.S. = Southern Europe; T. = Turkey. Reconstruction of Gondwana according to Courjault-Radé *et al.* (1992), position of Baltica after Torsvik *et al.* (1991), and Upper Silesia according to Moczydłowska (1995b).

associations from Upper Silesia are also strikingly similar to those from Iberia, particularly the Oville Formation. In the Iberian Peninsula acritarch assemblages, including numerous age-diagnostic taxa, are well recognized in different tectonostratigraphic zones. In the Cantabrian Mountains they occur most abundantly in the Lower Cambrian Herreria Formation and in the Middle Cambrian Oville Formation (Palacios & Vidal 1992). Comprehensive and important, yet largely unpublished, data from the Lower–Upper Cambrian 'Playon beds' in the Zafra area of Sierra Morena (Palacios 1993, and personal communication, 1997) provide additional evidence of a homogenous distribution of acritarch assemblages between Upper Silesia and Iberia.

Trilobite faunas in Baltica and Iberia differ substantially. Hence it is generally believed that during the Cambrian, each of these regions was part of a distinct faunal province, the Acado-Baltic and Gondwana provinces, respectively (McKerrow & Cocks 1995). The Acado-Baltic faunal province has been defined by occurrence of the olenellid trilobite taxa and encompasses Laurentia, Baltica and Avalonia, whereas the Gondwana province has been recognized by the presence of bigotinid trilobites in Morocco and Armorica (McKerrow & Cocks 1995). Bigotinid trilobites occur also in Siberia, and therefore it was suggested that both palaeocontinents were not distantly separated in Early Cambrian times (Pillola 1990). There is, however, inconsistency in the recognition of these faunal provinces and their palaeogeographic extension. According to palaeomagnetic data and proposed tectonic reconstructions, Avalonia and Armorica were marginal portions of Gondwana, yet they belonged to different provinces. This was explained by a postulated wide latitudinal separation of Avalonia from Armorica in Early Cambrian times, resulting in the development of differing faunas (McKerrow & Cocks 1995). However, some Cambrian trilobites regarded as Acado-Baltic occur also in Iberia and in Antarctica (whose location was on the opposite periphery of Gondwana; see below). Following alternative palaeogeographic reconstruction, Siberia was certainly more distantly located from Gondwana than Avalonia from Armorica (Dalziel 1995), nevertheless it shared bigotinid trilobites with Gondwana and Armorica.

Trilobites in Iberia are largely related to the Gondwanan faunal province and may be partly endemic. In spite of this relationship, a number of Lower and Middle Cambrian trilobite genera that are common among the Acado-Baltic faunas, such as *Strenuaeva, Kingaspis, Paradoxides* (*Eccaparadoxides*) and *Ellipsocephalus*, do also occur in Spain (Sdzuy 1971a, b; Liñán & Perejón 1981). *Parasolenopleura aculeata* and *Dolichometopus* sp. are Middle Cambrian taxa previously known only from the Acado-Baltic faunal province. Recently, they were recorded for the first time in Spain, providing new evidence that could suggest a closer connection between the Iberia and Baltica shelf areas than formerly envisaged (Liñán *et al.* 1995). A few Lower Cambrian trilobite taxa recovered in Upper Silesia are attributed to *Schmidtiellus panowi, Strenuaeva primaeva* and *Ellipsocephalus nordenskjoeldi*, and are age-diagnostic of the *Holmia kjerulfi* Zone (Orłowski 1975).

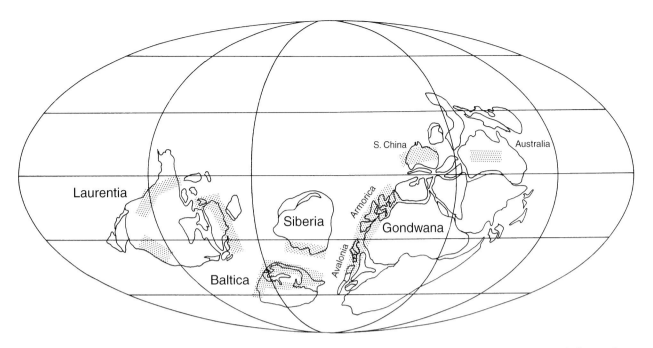

Fig. 19. Early Cambrian palaeobiogeographic reconstruction (modified after Courjault-Radé *et al.* 1992 and McKerrow & Cocks 1995) showing distribution of taxonomically comparable acritarch assemblages (shaded) in shelf basins extending into warm and temperate climatic belts. After Moczydłowska (1995b).

They belong to the Acado-Baltic faunal province, however, two of these genera are also known from Iberia.

The faunal migration, though environmentally controlled to a large extent, was also possible between different 'palaeoprovinces', as evidenced by the occurrence of olenellid trilobite taxa in Baltica, Upper Silesia, Iberia, and even in the Precordillera of Argentina. The latter occurrence might be explained by a recently inferred provenance of the Precordillera terrane from Laurentia. It has been suggested that the Precordillera terrane probably rifted apart from the area of the Gulf of Mexico (Astini *et al.* 1995 *in* Kerr 1995). This would obviously conform with the affiliation of the Precordillera terrane to the Acado-Baltic faunal province. An alternative explanation for this occurrence has been proposed, assuming possible migration of trilobites along contiguous shelves along margins of Gondwana, Baltica and Laurentia palaeocontinents in Cambrian times (Moczydłowska 1995b). This explanation is supported by the new record of the Early Cambrian trilobite genus *Kingaspis* in the Central Transantarctic Mountains (Palmer & Rowell 1995), showing its transprovincial and long-distance distribution. The genus *Kingaspis* has been previously recorded in Baltica (Poland, Lendzion 1978, Mens *et al.* 1990), Armorica (Spain, Sdzuy 1971a) and Gondwana (Israel, Parnes 1971; Morocco, Geyer 1990, Geyer *et al.* 1995).

Palaeogeographically the global acritarch records are most extensive for the Lower Cambrian as they comprise all recognized palaeocontinents (Fig. 19). Acritarch associations from South China and Australia consist of cosmopolitan taxa previously known from the Armorican and Avalonian margins of Gondwana, the Anabar shelf of Siberia, the Lublin and South Norwegian shelves of Baltica, and the Scotland, Svalbard and Greenland shelves of Laurentia. This is taken to indicate that all areas concerned might have been connected along the continental shelves, rendering possible phytoplankton dispersal. Interestingly, the most uniform global distribution of phytoplankton is observed in the Lower Cambrian *Holmia kjerulfi* Biozone (Baltica) and during at least partly time-equivalent interval, i.e. *Eoredlichia–Wutingaspis* (South China), *Bonnia–Olenellus* (Laurentia), *Callavia* (Avalonia) and Middle Tommotian (Siberia) (Moczydłowska & Vidal 1988; Moczydłowska 1991; Vidal & Peel 1993, Vidal *et al.* 1995; Zhuravlev 1995). This appears to coincide with maximum flooding and formation of nearshore phosphate deposits during the Early Cambrian sea level rise in *Holmia* times (Vidal & Moczydłowska 1996). The shelf areas with the record of these acritarch associations seem to occupy tropical and temperate latitudes (Fig. 19).

Middle–Upper Cambrian acritarch records derive from portions of Baltica, East and West Avalonia, and Armorica. During this period of time no significant taxonomic differentiation between acritarch associations from various areas can be distinguished. Numerous taxa are widespread and cosmopolitan in occurrence, whereas taxa regarded as endemic are few.

Because they are cysts of planktic protoctists, acritarchs have been thought to be poor palaeogeographic indicators. However, phytoplankton assemblages display a homogenous pattern of distribution and comprise numerous cosmopolitan species that occur throughout the Early Cambrian in Baltica, Avalonia, Armorica and Laurentia. This is consistent with the absence of effective palaeoenvironmental barriers and with a free dispersal of phytoplankton along the shelf areas concerned (Fig. 19). It has been inferred that pericratonic basins had been located along contiguous shelves facing the Iapetus Ocean and the Avalonian Seaway (Moczydłowska 1995b; Figs. 18 and 19 herein). The latter extended between the Avalonian and Armorican margins of Gondwana, the Finnmarkian shelf of Baltica and the Anabar shelf of Siberia (Fig. 18). In the reconstruction (Fig. 19), this narrow seaway was connected westwards with the Greenland and Svalbard shelves of Laurentia through the embryonic northern branch of the Iapetus Ocean. The contiguous shelves facing the Iapetus and Avalonian seaways were subsequently dismembered during the final stages of opening of the Iapetus Ocean and rifting apart of Avalonia from Gondwana in Ordovician times (McKerrow & Cocks 1995; Pharaoh *et al.* 1995).

Tectonic implications

The Cambrian terrigenous succession in Upper Silesia is gently dipping and unmetamorphosed, and it underlies the Lower Devonian Old Red Sandstone and a Middle Devonian – Lower Carboniferous carbonate platform association. This succession has been viewed as a platform cover overlying a folded and metamorphosed Upper Proterozoic sequence (Kotas 1973c, d, 1982a; Ślączka 1976, 1985c; Brochwicz-Lewiński *et al.* 1986), or as a succession deformed (folded and thrust) and metamorphosed in pre-Devonian times (Kowalczewski *et al.* 1984; Kowalczewski 1990, 1993). The latter hypothesis, advocating Caledonian deformation in Upper Silesia, now appears implausible. An alternative interpretation of the micropalaeontological data that were used initially to support the deformation hypothesis seems more likely (Moczydłowska 1993a, b, 1995a, 1996a), and new evidence is also provided by the present investigation. The hypothesis of Caledonian deformation was based on the erroneous interpretation of the stratigraphic ranges of the acritarch assemblages in Upper Silesia as indicating an inverted and repeated succession (Moczydłowska 1996a). This stratigraphic inversion is, however, only apparent, as the succession of acritarchs might be interpreted to indicate a normal succession of strata, also revealed by additional observations on the sedimentology and ichnofossils.

Rocks of the Sosnowiec and Goczałkowice successions (the Potrójna drillcore was not examined by the author) do not reveal cleavage, a feature that would be expected if the successions had undergone strong tectonic deforma-

tion as a result of recumbent folding (cf. Kowalczewski *et al.* 1984, Kowalczewski 1990). Neither was any sign of displacement observed of the presumed Cambrian portion of the succession within the Proterozoic tectonic breccia of the Czéchowice formation in the Goczałkowice drillcore (but see Kowalczewski *et al.* 1984; Kowalczewski 1990). The sedimentary rocks have undergone more diagenesis at the Sosnowiec site than at Goczałkowice, as shown by the varying grades of lithification of the rocks and the state of preservation of acritarchs and trace fossils. Neither succession exceeds the stage of diagenesis. However, the stage of mesocatagenesis has affected the rocks at Sosnowiec, whereas protocatagenesis affected those at Goczałkowice. These stages are inferred from the state of preservation and thermal maturation of particulate acid-resistant organic matter (see under 'Taphonomy'). The maturation of organic matter in the drillcore Potrójna is comparable to that in the Goczałkowice drillcore. Trace fossils, consisting of vertical and horizontal burrows, are well preserved and display no distortion (Figs. 7 and 9). They occur throughout the Sosnowiec and Goczałkowice successions (Figs. 6 and 8) in normal position on the bedding surface and as casts on bed soles. These features are in agreement with a normal sedimentary succession.

The observed normal stratigraphic succession of microfossils, normal position and state of preservation of trace fossils, low thermal maturation of organic-walled microfossils, and the lack of cleavage evidently indicate that the succession is undeformed and unmetamorphosed (Fig. 2).

Cambrian strata in the Potrójna borehole are underlain paraconformably by a sequence of polymictic, haematitic conglomerates (the Potrójna Formation, Fig. 10). The sequence overlies basement rocks with angular unconformity and an erosive contact, and was considered to be a remnant of the Late Proterozoic molasse of the Assynthian tectonic phase (Ślączka 1976, 1982, 1985b; Kotas 1982a). However, the inferred Late Proterozoic age of the molasse is in conflict with the inferred relationship to the Assynthian movements, since their age is Late Cambrian – Tremadocian (Sturt *et al.* 1980; Roberts & Sturt 1980; Brochwicz-Lewiński *et al.* 1981; Coward & Potts 1985). The conglomerates extend through the southeastern part of Upper Silesia (the Potrójna IG-1, Piotrowice 1, Raciborsk 2 and Łapczyca 2 boreholes; Ślączka 1985b). Lithologically similar deposits of the Pszczyna member occur in the Goczałkowice IG-1 borehole (Fig. 8) and were interpreted as the basal conglomerate of the transgressive Cambrian sequence (Kotas 1982b, Kowalczewski 1990). However, in the latter succession the contact of the Pszczyna member with the overlying sandstones was not cored, and it occurs in a fault zone recognized by geophysical methods (Kowalczewski 1990). Thus, whether it belongs to the transgressive sequence is not clear. The relative age of these red beds can be indirectly inferred to be

Late Neoproterozoic (probably Vendian) from their stratigraphic position in respect to overlying fossiliferous Early Cambrian strata (equivalent to the *Schmidtiellus* Zone), above an intervening unfossiliferous sandstone sequence (that might belong to the *Platysolenites* Zone) (Fig. 8). It appears feasible that both the Potrójna Formation and the Pszczyna member are coeval. Assuming that the red beds of the Potrójna Formation represent a true molasse sequence, as suggested (Ślączka 1985b), this would be the post-Cadomian molasse in the Upper Silesia terrane (Moczydłowska 1997).

Basement rocks underlying the angular unconformity consist of folded metasediments affected by greenschist facies metamorphism (the Skawa Formation, Fig. 10, and the Czéchowice Formation, Fig. 8), and plutonic and high-grade metamorphic rocks (Kotas 1982a; Ślączka 1985c; Moczydłowska 1996a). Deformation and metamorphism of the metasedimentary sequence is evidently pre-Early Cambrian and is attributed to the Cadomian tectonothermal event (Brochwicz-Lewiński *et al.* 1981, 1983, 1986; Moczydłowska 1995a, 1996a). The crystalline basement complexes (gabbros, granitoids, schists, gneisses and amphibolites) underlie the metasediments (Kotas 1982a), but the tectonostratigraphic relationships between various igneous and high-grade metamorphic rocks have not been established with certainty, and no isotopic ages are available (Moczydłowska 1996a). Structurally, the crystalline and metasedimentary basement complexes were referred to the Proterozoic or Cadomian Upper Silesia Massif (Kotas 1982a; Bukowy 1982; Brochwicz-Lewiński *et al.* 1981, 1986).

In the close vicinity to Upper Silesia, the Cadomian basement rocks are recognized in the Brno Massif in the Moravian area (Brochwicz-Lewiński *et al.* 1981; van Breemen *et al.* 1988; Oliver *et al.* 1993; Franke 1995) and in the Bohemian Massif (Franke 1992). Further afield in Europe, such basement complexes occur in the Lüneberg Massif (or North German Massif and Lusatia, Erdtman 1991), the Midlands Massif (Pharaoh *et al.* 1987; Berthelsen 1992a), the Brabant Massif (Woodcock 1991), in the type area of the Armorican Massif and in the Iberian Massif (D'Lemos *et al.* 1990; Dallmeyer & Martinez Garcia 1990). The crustal blocks comprising the Midlands, Brabant and Lüneberg Massifs form the Eastern Avalonian Terranes that were accreted to Baltica in Late Ordovician (Pharaoh *et al.* 1995) or Early Silurian times (McKerrow & Cocks 1995) along the Trans-European Fault, presumably coinciding with the Avalonia–Baltica suture (Berthelsen 1992a). The extension of Cadomian basement from the Lüneberg Massif, deeply buried under the Variscan thrust belt around the Bohemian Massif, has been expected to reappear in the 'Silesian massif' (Berthelsen 1992b). This seems to be true, as in addition to the Cadomian deformation and metamorphic events recorded in the upper segment of the base-

ment in Upper Silesia, there are other circumstances suggesting affinity of the Upper Silesia block with the Eastern Avalonian Terranes.

In the Midlands and Lüneberg Massifs the consolidated Cadomian basement is overlain, with sub-Cambrian unconformities, by relatively undisturbed Early Palaeozoic sedimentary covers (Pharaoh *et al.* 1987, Berthelsen 1992a, Pharaoh *et al.* 1995). Comparably, in the Upper Silesia terrane almost flat-lying Cambrian strata (there are no Ordovician or Silurian rocks in the area; Bukowy 1982) form a platform cover on basement rocks. Here, the erosive surface of the basement is additionally overlain with angular unconformity by possible remnants of the post-Cadomian molasse (presumably Vendian) underlying the paraconformably succeeding marine transgressive Cambrian strata. The Midlands and Lüneberg Massifs form the border zone against a belt of Caledonian deformation that is the Avalonian accretionary belt (Berthelsen 1992a). The rocks within the belt were deformed during the closure and compression of the Tornquist sea in Ordovician–Silurian times (the Caledonian orogeny) caused by the collision between East Avalonia and Baltica. The East Avalonia – Baltica suture, indicated by the Trans-European Fault, is recognized across the North Sea and North Germany, and approximately in the Małopolska area in southern Poland (Berthelsen 1992a, 1993). The Upper Silesia terrane forms a crustal wedge incised between the areas of Caledonian deformation (e.g., the Holy Cross Mountains, the Małopolska area and the Krakowian Belt, and the Sudetes Mountains) along the Moravian–Silesian fracture and the Kraków–Myszków Fault Zone (Oliver *et al.* 1993; Johnston *et al.* 1994; Moczydłowska 1995a, 1996a, Fig. 2).

In the area of the Holy Cross Mountains, Małopolska and the Krakowian Belt (i.e. area around the Kraków–Myszków Fault Zone), two major Caledonian tectonic events are recorded by regional angular unconformities, e.g., Cambrian – Lower Ordovician and Silurian – Lower Devonian (Jurkiewicz 1975; Harańczyk 1982, 1994; Brochwicz-Lewiński *et al.* 1986). The time constraint on Early Caledonian deformation in the Małopolska area (the Nida Trough) is provided by a recent U–Pb dating on zircons from tuffs within a sedimentary succession of the Książ Wielki Formation. The succession is tectonically deformed, weakly metamorphosed into greenschist facies and underlies the sub-Lower Ordovician angular unconformity in the Książ Wielki IG-1 borehole (Jurkiewicz 1975; Kowalczewski 1981). The numerical age of the tuffs is 549±3 Ma and indicates a latest Vendian age for the Książ Wielki Formation (Compston *et al.* 1995). Thus far, this formation is the oldest documented sedimentary succession with a well-established age recorded in the Małopolska area, and the tectonothermal event that caused its deformation is evidently Cambrian. Basement rocks underlying the Książ Wielki Formation are unknown

(Jurkiewicz 1975). In the Holy Cross Mountains, Precambrian rocks are not documented and the oldest strata palaeontologically recognized are of Early Cambrian age. There, basement complexes have never been reached by boreholes. The area of the Caledonian deformation in the Holy Cross Mountains, Małopolska and the Krakowian Belt, underlying basement of unknown age and provenance, is flanked by the Baltica craton and by the Upper Silesia terrane with Cadomian, East Avalonia-derived basement (Fig. 1). As such, this area obviously belongs to the Avalonian accretionary belt and embraces the Trans-European Fault.

In the Sudetes Mountains, Caledonian deformation, metamorphism and emplacement of igneous rocks, though greatly obscured by subsequent Hercynian deformation, has been recently reviewed and documented by new U–Pb datings (Oliver *et al.* 1993 and Johnston *et al.* 1994, but see also Oberc 1977, 1986; Don & Żeláźniewicz 1990, Aleksandrowski 1994, Żeláźniewicz & Franke 1994, Moczydłowska 1996a, Johnston *et al.* 1996). The new interpretation of the tectonostratigraphic relationships between various terranes amalgamated along the Intra-Sudetic Fault enables this crustal fracture to be recognized as the Tornquist Suture, marking the closure of the Tornquist Sea during the Caledonian orogeny (Johnston *et al.* 1994). Based on the recognition of different terranes in southern Poland and the distribution of crustal fractures that originated contemporaneously and which may be related to the Tornquist Suture recognized in the Sudetes, the Kraków–Myszków Fault Zone was indicated as the possible extension of this suture (Moczydłowska 1996a, 1997). The Caledonian folding, thrusting, strike-slip displacement, transpressional faulting and the shear zone recognized in the Kraków–Myszków Fault Zone (Brochwicz-Lewiński *et al.* 1981, 1983; Harańczyk 1982, 1994; Żaba 1994, 1995) are most pronounced within the accretionary belt between the Upper Silesia terrane and the Teisseyre–Tornquist Zone (Fig. 2). This fracture may therefore correspond to the Tornquist Suture. A palaeogeographic reconstruction based on palaeomagnetic data, climatic belts, facies development and fossil distribution during Cambrian times (Torsvik *et al.* 1991; Courjault-Radé *et al.* 1992; McKerrow & Cocks 1995), in which the Upper Silesia terrane is recognized as a distal segment of the East Avalonia microplate (Moczydłowska 1995b; Fig. 18), is in agreement with this hypothesis.

In brief, the tectonic history of the Upper Silesia terrane may be summarized as follows:

The final consolidation of the crustal block forming the basement in Upper Silesia occurred during the Late Cadomian tectonothermal events, identified by deformation and metamorphism that predates the Cambrian. Upper Silesia was presumably part of the destructive margin of the Gondwana palaeocontinent, along with other Avalonian and Armorican blocks. Subsequently, it was uplifted and denuded as shown by the erosive top surface of the metasedimentary sequence (the Czéchowice Formation, Goczałkowice borehole). By the end of the Vendian, post-Cadomian molasse had accumulated (the Pszczyna member, Goczałkowice borehole). The extensional tectonics and rifting at the margin of Gondwana resulted in subsidence and marine transgression over the Upper Silesia area in Early Cambrian times (the Goczałkowice formation). Siliciclastic sedimentation persisted throughout Cambrian with probably short-lasting shallowing-up episodes and sedimentation breaks (the Goczałkowice, Sosnowiec and Jaszczurowa formations). As a part of East Avalonia, Upper Silesia occupied an intermediate position between the margins of Gondwana and Baltica during Cambrian times and experienced the same tectonic history. The Avalonian microplate was rifting apart from Gondwana throughout Cambrian times (Pharaoh *et al.* 1995), had drifted away by the Early Ordovician (Arenig; Scotese & McKerrow 1990, McKerrow & Cocks 1995), and collided with Baltica in the Late Ordovician (Pharaoh *et al.* 1995) or Early Silurian (McKerrow & Cocks 1995). The absence of Ordovician and Silurian rocks in Upper Silesia (Bukowy 1982) might be interpreted as being the probable result of denudation. Since Early Devonian times, the Upper Silesia terrane has undergone comparable sedimentary development and shares a common tectonic history with the Baltica palaeocontinent.

Systematic palaeontology

Taxonomic comments

The taxonomy of acritarchs is based entirely on morphological features. Insufficient knowledge of their biology and evolutionary affinities has resulted in an informal taxonomy. This approach was adopted at an early stage in investigation of these microfossils (Evitt 1963; Downie *et al.* 1963; Staplin *et al.* 1965; Eisenack *et al.* 1973; Tappan 1980) and is the only practical way of dealing with them, in so far as information concerning the structure of the wall and chemical composition of its polymers is not generally available. This could possibly allow establishment of the relationships between various groups of acritarchs and their living counterparts. Hitherto, biochemical and structural studies (Jux 1968, 1969a, 1969b, 1971; Kjell-ström 1968; Colbath 1983; Amard 1992) have dealt only with a few selected taxa and have not provided conclusive evidence of their relationships. Obviously, a limited array of morphological features could point towards a poly-phyletic group of protoctists, producing resistant sporopollenin-like encysted life stages (Evitt 1963; Tappan 1980).

Many new records of morphologically complex and variable acritarchs have been obtained in recent years from the Lower Palaeozoic and Neoproterozoic (Vanguestaine & Van Looy 1983; Playford & Martin 1984; Turner 1984; Pittau 1985; Welsch 1986; Knoll & Swett 1987; Bagnoli *et al.* 1988; Martin & Dean 1988; Playford & Wicander 1988; Baudet *et al.* 1989; Hagenfeldt 1989a, b; Le Hérissé 1989; Zang & Walter 1989, 1992; Colbath 1990; Tongiorgi & Ribecai 1990; Volkova 1990; Albani *et al.* 1991; Di Milia 1991; Moczydłowska 1991; Vanguestaine 1991; Martin 1992; Zang 1992; Mendelson 1993; Vidal & Peel 1993; Moczydłowska *et al.* 1993; Palacios & Vidal 1992; Yin Lei-ming 1994; Butterfield *et al.* 1994; Young *et al.* 1994). Morphological diversity among acritarchs is being recognized increasing and this demands adjustment of the initial taxonomic concepts and greater precision in the diagnoses of form-genera and form-species. This is reflected in a recurring process of taxonomic revision, emendation and recognition of new taxa that attempts to accommodate these needs. Recently, more attention has been paid to the fine morphological details of acritarchs, e.g., the morphology of processes, being either solid or hollow, their connection or lack thereof with the interior of the vesicle, their abundance and pattern of distribution, and the ornamentation of the vesicle surface and processes (Tappan and Loeblich 1971, Loeblich & Tappan 1978, Wicander 1974, Colbath 1990, Volkova 1990, Moczydłowska 1991, Vidal & Peel 1993). The use of arbitrary size limits in diagnoses cannot be strictly accepted, because there are many examples of specimens with identical morphologies to defined species but which differ in being beyond any selected size limits. By comparison with modern phytoplankton taxa, demonstrating broad range of dimensions within discrete species, a sharp size limit has neither biological nor taxonomic value (Loeblich 1970; Colbath 1979; Moczydłowska 1991). However, an approximate size range is used in practice to distinguish fossil genera and species. In contrast, some genera are still treated as a 'waste basket' collecting microfossils of various morphologies, such as *Micrhystridium* Deflandre 1937 and *Comasphaeridium* Staplin, Jansonius & Pocock 1965, with added arbitrarily chosen limits in the number of processes and their dimensions (Sarjeant & Stancliffe 1994). A recent emendation of *Micrhystridium* Deflandre 1937 by Sarjeant & Stancliffe (1994, p. 12) retains the former circumscription of the genus as 'polymorphic', thus including any acanthomorphic acritarchs of small

size, regardless of whether the processes are solid or hollow and connected with or separated from the vesicle cavity. The nature of the vesicle wall, either single- or double-layered, psilate or ornamented, is also neglected, and only the number of processes, established as 9–35, is considered to be a diagnostic feature. The reason for choosing these numbers is nowhere stated; nor is it stated whether the number of processes should be counted around the outline of the vesicle or on its entire surface. An additional and enigmatic feature proposed in this emendation is that a few spines may have small holes (sic!) in their mid section, a feature that has no precedent among the previously defined diagnostic characters of acritarchs. Sarjeant & Stancliffe (1994) also erected subgenera among micrhystrids, formerly recognized as discrete genera of Cretaceous picoplankton (Habib & Knapp 1982), and subspecific entities, e.g., forma. This procedure adds two additional taxonomic categories for identification of the microfossils. However, Sarjeant & Stancliffe (1994) indicated uncertainty as for the relationship between the type subgenus *Micrhystridium Micrhystridium* and other subgenera.

A complete review of the taxonomic re-evaluation by Sarjeant & Stancliffe (1994) is outside the scope of this paper. However, some specific comments are pertinent. Sarjeant & Stancliffe (1994) discussed earlier taxonomic proposals by Loeblich (1970), Wicander (1974), Colbath (1990), Habib & Knapp (1982), Eisenack *et al.* (1979a, 1979b) and Moczydłowska (1991), introducing various misconceptions based on misunderstandings of the diagnoses of certain genera and erroneous citations. Thus, Sarjeant & Stancliffe (1994, p. 4) misquoted Moczydłowska (1991, p. 46) by assembling together portions of text referring to the discrete genera *Asteridium* and *Heliosphaeridium*. They further erroneously concluded that these genera were distinguished from *Micrhystridium* by age assignment alone (Sarjeant & Stancliffe 1994, p. 4). In fact, Cambrian microfossils were excluded from the genus *Micrhystridium* not because of the fact that they might be of different affinity, but because of the inadequate diagnosis (original and emended) leading to inconsistencies associated with the taxonomy of microfossils attributed to this genus, including various morphotypes ('polymorphism') (Moczydłowska 1991, pp. 44–46). Acritarchs attributed to *Asteridium* and *Heliosphaeridium* possess solid and hollow processes, respectively (Moczydłowska 1991, p. 47 and 58), features that were not accepted as diagnostic by Sarjeant & Stancliffe (1994). These features contrast with those applying to the polymorphic genus *Micrhystridium*. Re-evaluation of type collections of younger forms of *Micrhystridium* is needed, and the emendation by Sarjeant & Stancliffe (1994), maintaining their 'polymorphism' and adding a limit to the number of processes (9–35), has not added much order to the existing taxonomic chaos.

Sarjeant & Stancliffe (1994, p. 4) additionally considered that both *Asteridium* Moczydłowska 1991 and *Heliosphaerdium* Moczydłowska 1991 overlap the diagnosis of *Eomicrhystridium* Deflandre 1968. However, they did not consider the probability that the type-species of *Eomicrhystridium*, *E. barghoornii* (Deflandre 1968, Pl. 1:1–3), represents either badly degraded sphaeromorphic acritarch (Mendelson & Schopf 1992, p. 886) or a fortuitously shaped fragment of particulate kerogen (personal examination by Vidal, personal communication, 1994). Thus the genus *Eomicrhystridium* refers to an misidentified object, probably not a true microfossil. The other species of the genus, *E. aremoricanum* Deflandre 1968 (Deflandre 1968, Pl. 1:4–5), is either a poorly preserved coccoidal cyanobacterium (Mendelson & Schopf 1992, p. 928) or degraded particulate kerogen (my interpretation) and, as such, certainly does not possess any processes. In this context the description by Sarjeant & Stancliffe (1994, p. 28) of a variety of process shapes (slender, capitate spines, massive pyramids) in both species of *Eomicrhystridium* and a comparison to dinoflagellate cysts is puzzling. Other microfossils attributed to *Eomicrhystridium* have subsequently been revised. Thus, specimens of *E. barghoornii* described by Hofmann (1971) from the same bed as the alleged type specimen (Gunflint Formation in Ontario) were reinterpreted as a distorted specimen of *Huroniospora* (Moore *et al.* 1992, p. 234) or *Kakabekia* sp. (Mendelson & Schopf 1992, p. 886). *Eomicrhystridium*? sp. reported from the Neoproterozoic Serie Negra in Spain (Gonçálves & Palacios 1984, Liñán & Palacios 1987) is a poorly preserved sphaeromorphic acritarch, affected by pyrite growth with the appearance of triangular pseudoprocesses (Palacios 1989, p. 11; Mendelson & Schopf 1992, p. 929). Furthermore, examination of samples from the type locality of *Eomicrhystridium aremoricanum* (Brioverian in Ville-au-Roi-en-Maroué, France; Deflandre 1968) revealed a comparable state of preservation and thermal alteration of the organic matter as observed in the Serie Negra, undoubtedly owing to the same grade of metamorphism affecting rocks in both areas (Palacios 1989, p. 11).

Recognizing the morphology of the processes as a diagnostic trait, the genus *Globosphaerdium* Moczydłowska, 1991, possessing solid processes, was distinguished from *Baltisphaeridium* Eisenack, 1958b, emend. Eisenack, 1969, having hollow processes (Moczydłowska 1991, pp. 46, 54). Contrary to Sarjeant & Stancliffe's assertions (1994, p. 23), *Globosphaeridium* was compared with *Comasphaeridium* by Moczydłowska (1991, p. 46, 54), though comparison with *Filisphaeridium* was thought to be unnecessary, the latter having obviously distinctive features (cylindrical processes not tapering as *Globosphaeridium*, and capitate processes tips, not acuminate). Sarjeant & Stancliffe (1994, p. 23) attributed the perceived absence

of comparisons by Moczydłowska (1991) to 'her philosophy' that Cambrian species merited generic differentiation from later species, which would indeed have been a less constructive approach.

Sarjeant & Stancliffe (1994, p. 25) treated *Comasphaeridium*, in contradiction to the diagnosis, as another 'polymorphic' genus which they regarded as synonymous with the genera *Elektoriskos* Loeblich 1970, *Globosphaeridium* and *Heliosphaeridium*. Among the diagnostic features of *Comasphaeridium* are densely crowded, thin, solid, simple, flexible, hair-like spines (Staplin *et al.* 1965, p. 192). The three synonymized genera are significantly different, having well-spread or even rare processes, more stiff than flexible. Additionally, *Heliosphaeridium* possesses hollow processes. In their emendation of *Comasphaeridium*, although the original diagnosis by Staplin *et al.* (1965) is concise and well understood, Sarjeant & Stancliffe (1994, p. 25) added little substance but arbitrarily chose a number of processes (more than 35) and a size limit of the latter (over 25% of the body diameter). These limitations are artificial and meaningless for the identification of microfossils. Even if they are accepted, all the genera synonymized with *Comasphaeridium* usually possess fewer processes, and *Globosphaeridium* definitely has shorter processes. Sarjeant & Stancliffe's emendation of *Comasphaeridium* is thus not followed here.

A different approach was applied by Sarjeant & Stancliffe (1994, p. 67) to species of *Asteridium*, by scattering them over three genera: *Micrhystridium*, *Comasphaeridium* and *Filisphaeridium* Staplin *et al.*, 1965. Even ignoring the taxonomically problematic *Micrhystridium*, this is a remarkable proposition since none of the mentioned species, *Asteridium lanatum*, *A. pallidum*, *A. spinosum* and *A. tornatum*, display features diagnostic of *Comasphaeridium* or *Filisphaeridium* (except for having solid processes). Similarly, it was proposed that species of *Heliosphaeridium* Moczydłowska 1991 (with hollow processes as a diagnostic feature) should be attributed to *Micrhystridium* (hollow, solid or half-solid processes) and *Comasphaeridium* (exclusively solid processes). Most of the species were 'returned' to *Micrhystridium* (Sarjeant & Stancliffe 1994, p. 70), despite the fact that its generic diagnosis is still vague. However, the type species of *Heliosphaeridium*, *H. dissimilare* (Volkova 1969b) Moczydłowska 1991, having hollow and not particularly numerous processes, was 'unequivocally' transferred to *Comasphaeridium* (whose diagnosis includes solid and very abundant processes).

The many misunderstandings and misquotations in Sarjeant & Stancliffe's (1994) paper (there are a number of them, in addition to the ones discussed above) and the apparent lack of examinations of collections other than Mesozoic ones make their taxonomic revision impossible to follow.

Review of selected genera

Cambrian acanthomorphic acritarchs possessing solid processes are attributed to the form genera *Asteridium* Moczydłowska, 1991, *Comasphaeridium* Staplin *et al.*, 1965, and *Globosphaeridium* Moczydłowska, 1991, whereas those having hollow processes are referred to *Baltisphaeridium* Eisenack, 1958b, emend. Eisenack, 1969, *Skiagia* Downie, 1982 emend. Moczydłowska, 1991, *Heliosphaeridium* Moczydłowska, 1991, *Multiplicisphaeridium* Staplin, 1961 restricted by Staplin *et al.*, 1965, emend. Eisenack, 1969, emend. Lister, 1970, *Solisphaeridium* Staplin *et al.*, 1965, restricted herein, *Trichosphaeridium* Timofeev, 1966, and *Celtiberium* Fombella, 1977. Individual genera within these two morphological groupings are distinguished by the shape of the processes, their abundance and dimensions, and the overall shape and diameter of the vesicle. An additional diagnostic feature in the second group of taxa is the communication, or lack thereof, between the process cavity and the vesicle. Most of the above genera are recognized on the basis of the original diagnoses with some subsequent emendations. The taxonomic concept of *Solisphaeridium*, *Trichosphaeridium*, *Celtiberium*, *Polygonium* and the non-acanthomorphic genus *Adara* are discussed below.

Baltisphaeridium and *Trichosphaeridium* were not recorded in the present study. Both genera are known from the Cambrian (Timofeev 1959, 1966; Bagnoli *et al.* 1988; Volkova 1990; Fensome *et al.* 1990; Moczydłowska 1991; Moczydłowska & Crimes 1995), but thus far only specimens described from its upper series may be convincingly attributed to the genera. They are mentioned here to elucidate the taxonomic ideas developed around acanthomorphic taxa among which *Baltisphaeridium* played the primary role. The two genera might be superficially similar in the shape of the processes, but they are easily discriminated by the free communication between the process cavity of the vesicle in *Trichosphaeridium* (see below), as opposed to the separated process and vesicle cavities in *Baltisphaeridium*.

Among taxa with solid processes *Comasphaeridium* is characterized by very abundant and tightly arranged, flexible and tapering processes. *Asteridium* and *Globosphaeridium*, sharing similar shapes of processes (simple, with acuminate, blunt or rounded tips), are distinguished by the size of the vesicle, being minute to small (a few micrometers to around 20–25 μm) in *Asteridium* and medium to big (around 20 μm to 50–60 μm) in *Globosphaeridium*. However, there is no sharp size limit between the two genera, and their recognition might be subjective since the range of dimensions overlaps.

Within the group of taxa with hollow processes, separated from the vesicle cavity by the vesicle wall or a plug, are *Baltisphaeridium* and *Skiagia* (Eisenack 1969; Downie 1982; Moczydłowska 1991). The generic diagnostic feature defining species of *Skiagia* are funnel-like distal process terminations. The distal portion of the processes in *Baltisphaeridium* are tapering and acuminate, occasionally dichotomizing (Eisenack 1969, p. 249). Thus the two genera are recognized by the morphology of the process terminations alone.

The common diagnostic feature of hollow processes that freely communicate with the vesicle interior is shared by *Heliosphaeridium*, *Multiplicisphaeridium*, *Solisphaeridium*, *Trichosphaeridium* and *Celtiberium*. *Heliosphaeridium* is characterized by its minute to small diameter of the vesicle and simple processes with acuminate, flared or dichotomizing tips. *Multiplicisphaeridium* has multifurcate process terminations. *Solisphaeridium* and *Trichosphaeridium* have simple processes, but *Trichosphaeridium* is distinguished by an abundant and dense cover of slim processes. *Solisphaeridium* is reviewed below in more detail. *Celtiberium* is easy to recognize among other genera because of the unique appearance of the wide processes, like rigid or turgid fingers.

The morphology of the processes in *Trichosphaeridium* was described as 'hair-like' by Timofeev (1959, 1966), but no further details nor photomicrographs of the species were provided. Volkova (1990) attributed specimens with clearly hollow processes (Volkova 1990, Pl. 20:5) to this genus, thus recognizing this particular feature as a characteristic of *Trichosphaeridium*. This was followed by Moczydłowska & Crimes (1995), whereas Sarjeant & Stancliffe (1994, p. 28) considered *Trichosphaeridium* Timofeev 1966 to be a junior synonym of *Filisphaeridium*. This proposed synonymy followed their emendation of *Filisphaeridium* to include processes with acuminate terminations as well as those with differentiated terminations. They further concluded that *Trichosphaeridium* differs from *Filisphaeridium* only by its relative large size. However, the processes of *Trichosphaeridium* (which are simple, occasionally dichotomizing and have acuminate tips) are hollow, though very thin. This feature contrasts with the solid processes that are diagnostic for *Filisphaeridium*.

The general habit of *Celtiberium* Fombella, 1977, is conspicuously different from that of *Polygonium* Vavrdová, 1966. In *Celtiberium* (Fombella 1977, p. 117; Fombella 1978, pp. 250–251) the processes are wide and cylindrical with rounded terminations, in contrast to the simple and tapering processes of *Polygonium* (Vavrdová 1966, p. 413). The common features of the two genera are the free connection between the process and vesicle cavities and the broad bases of the processes. Sarjeant & Stancliffe (1994, p. 42) regarded *Celtiberium* as a junior synonym of *Polygonium*, thus rejecting it and by default transferring all its species to *Polygonium* as new combinations. This proposition is inconsistent with the diagnoses of the latter genus, both as originally defined or as later emended by Sarjeant & Stancliffe. In their emendation of *Polygonium*, they emphasized (Sarjeant & Stancliffe 1994, p. 43) that the processes are distally acuminate, a feature that is discrepant with the diagnosed and clearly observable rounded process terminations of *Celtiberium*. The genus *Celtiberium* Fombella 1977 is retained as a separate taxon. Acritarchs of this genus differ from *Adara* by having fewer, more widely separated and longer processes. The genus includes *Celtiberium geminum* Fombella 1977, *C. clarum* Fombella 1978, *C. dedalinum* Fombella 1978, and *C.? papillatum* n.sp. The species *Celtiberium ondulatum* Fombella 1979, a *nomen nudum*, is synonymous with *Adara undulata* n.sp. described herein.

The genus *Solisphaeridium* Staplin *et al.* 1965 was erected to group acanthomorphic acritarchs that differ from *Micrhystridium* by having subcylindrical spines that make a sharp angle with the vesicle wall and have narrow process bases. It differs from *Baltisphaeridium* because the processes are open to the vesicle cavity (Staplin *et al.* 1965, pp. 183–184). The morphology of the processes, solid or hollow, was not considered as a significant diagnostic feature in the original concept of the genus. This feature is, however, presently used to define separate genera and, if not defined, the general characters of *Solisphaeridium* would nevertheless coincide with those of *Micrhystridium*.

The genus *Solisphaeridium* encompasses microfossils with variable process morphology that can be hollow, semi-hollow or solid, and whose cavities, if present, communicate with the vesicle interior (Staplin *et al.* 1965, pp. 183–184, Text-fig. 3 and 10). *Solisphaeridium stimuliferum* (Deflandre 1938) Staplin *et al.* 1965 was selected as the type species (Staplin *et al.* 1965, pp. 183–184, Pl. 18:1–2), but without further information about which kind of processes it possesses. This combination of species was based on the *Hystrichosphaeridium stimuliferum* Deflandre 1938 (Deflandre 1938, p. 192, Pl. 10:10) (=*Micrhystridium stimuliferum* in Staplin *et al.* 1965, p. 183), subsequently transferred to *Baltisphaeridium* and then to *Solisphaeridium* (complete synonymy in Sarjeant 1968, p. 223). The species illustrated by Staplin *et al.* (1965, Pl. 18:1–2) seems to have hollow processes and a very thick, firm vesicle wall, but it is uncertain whether the cavity of the processes communicates with the vesicle interior. Sarjeant (1968, p. 222) emended the diagnosis of the genus *Solisphaeridium* by adding the size limit of the mean vesicle diameter, being greater than 20 μm, and the presence of a median split type of opening, but he accepted the variable morphology of the processes. Nevertheless, in remarks on the type species, he indicated the presence of spine cavities communicating with the vesicle interior (Sarjeant 1968, p. 223, Fig. 1, Pl. 3:6). This observation was based on the examination of the holotype and material collected from the type locality. The wall of the vesicle in the illustrated specimen (Sarjeant 1968, Pl. 3:6) seems also to be thick.

Pocock (1972, p. 113) attributed *Solisphaeridium stimuliferum* (Deflandre) n.comb. as a new combination to the type species of *Solisphaeridium* (though it had already been recombined as such by Staplin *et al.* 1965) and ascribed to it solid processes (Pocock 1972, p. 113, Pl. 28:21–23). This feature was shown only on one specimen (Pocock 1972, Pl. 28:229) that might be of a different species, whereas two other specimens (Pocock 1972, Pl. 28:21, 23) display hollow processes and are similar to the micrographs of the species published by Staplin *et al.* (1965) and Sarjeant (1968). Wicander (1974) and Colbath (1990) considered the communication of hollow processes with the vesicle cavity to be a diagnostic feature of *Solisphaeridium,* and all species attributed to this genus described by them from Devonian strata display this feature. Similarly, Playford & Martin (1984) attributed their new species to *Solisphaeridium*, though only the proximal part of the processes are hollow and communicate with the vesicle cavity, the distal portion being very thin and tapering. Fensome *et al.* (1990) recognized the genus *Solisphaeridium* Staplin *et al.*, 1965 emend. Sarjeant, 1968 (thus possessing variable processes, hollow or solid), but retained *S. stimuliferum* (Deflandre, 1938) emend. Pocock, 1972 (possessing, according to the diagnosis, solid processes), as the type species. This would imply that the genus *Solisphaeridium* has exclusively solid processes, but this is inconsistent with the diagnosis of the genus and the observations on the type species by Sarjeant (1968).

Sarjeant & Stancliffe (1994, p. 12) considered *Solisphaeridium* as a junior synonym of *Micrhystridium* Deflandre, 1937, newly emended by them. However, by their diagnosis, only specimens of *Solisphaeridium* with a vesicle diameter less than 20 μm (with some exceptions) could be included in *Micrhystridium*. Sarjeant's (1968) earlier view of solisphaerids having a larger vesicle diameter than micrhystrids (although a number of species of both genera have overlapping dimensions) is, however, shared by a number of researchers. Species have been attributed to *Solisphaeridium* that have a range of vesicle diameter of 20–30 μm (Slaviková 1968), 17–29 μm (Wicander 1974), 19–33 μm (Playford & Martin 1984), 16–29 μm (Eklund 1990), and 20–36 μm but also 14–25 μm (Colbath 1990). I propose that *Solisphaeridium* Staplin *et al.* 1965 be restricted to acritarchs having simple, unbranching, hollow processes that are connected to the vesicle cavity, and with a vesicle that is generally larger in diameter than *Heliosphaeridium* Moczydłowska, 1991 (the latter also bearing hollow processes). Both genera may have slightly overlapping vesicle diameters. The monomorphic *Solisphaeridium* as restricted here cannot be synonymized with 'polymorphic' *Micrhystridium* that includes a broad array of morphotypes and is thus poorly defined, particularly as species with hollow processes have been excluded from *Micrhystridium* and attributed to *Heliosphaeridium* (Moczydłowska 1991).

An additional genus frequently occurring in Cambrian strata, *Adara*, also bears hollow, but short and wide processes, and the overall shape of the vesicle differs from acanthomorphic taxa. It also differs from *Polygonium*, despite sharing features in common (such as wide conical bases of the hollow processes and a polygonal outline of the vesicle), by having conspicuously shorter processes with rounded or blunt distal portions. *Polygonium* is a distinctive genus, primarily because of the polygonal vesicle outline. The latter genus includes as a junior synonym *Goniosphaeridium* Eisenack, 1969, emend. Kjellström, 1971, emend. Turner, 1984 (Le Hérissé 1989, p. 181; Albani 1989; Fensome *et al.* 1990, p. 232 and 405; Sarjeant & Stancliffe 1994, p. 68; Moczydłowska & Crimes 1995).

The genus *Adara* Fombella, 1977, emend. Martin, 1981 (in Martin & Dean 1981), groups acritarchs with spheroidal vesicles bearing numerous stout and short, conical processes that have rounded terminations, are hollow, and communicate with the cavity of the vesicle. Because of the abundance of the processes, their wide bases are in very close proximity or grade into one another, forming a wavy outline on the vesicle (compare Fombella 1977, Pl. 1:16, Fig. 1:6; Fombella 1978, Pl. 1:16–17; Martin *in* Martin & Dean 1981, Pl. 1:20–22, Pl. 4:7, 9–10). These features are distinct from those defining the genus *Buedingiisphaeridium* Schaarschmidt, 1963 (Schaarschmidt 1963, p. 69–70, Pl. 20:4–6, Text-fig. 26), which has small and massive processes scattered on the surface of the wall and being well separated. These morphologic elements are here regarded as sculptural elements of the vesicle wall, rather than being true processes. Nevertheless, the two genera have also been considered synonymous, with priority given to the latter.

Cramer & Diez (1979) referred to the type species of *Adara* (*A. matutina* Fombella 1977) as '*Buedingiisphaeridium matutinum* (Fombella 1977)', '*B. matutina* (Fombella 1977)', and '*Lophosphaeridium matutinum* (Fombella 1977)' (Cramer & Diez 1979, pp. 43, 45, 51). Fombella (1979, 1982) used the epithet '*Buedingiisphaeridium matutinum* (Fombella 1977)'. However, none of these recombinations were validly published, as they were neither identified as new combinations nor followed by any comments or references. The subsequent emendation of *Buedingiisphaeridium* by Sarjeant & Stancliffe (1994, p. 24), in which they combined the presence of solid, hollow or partially hollow ornaments, results from the fusion of two distinctive genera (*Buedingiisphaeridium* and *Adara*). This emendation was based on the earlier description of *Buedingiisphaeridium* by Staplin *et al.* (1965). Staplin *et al.* (1965, pp. 179–180) added as a diagnostic feature of the genus, to the originally diagnosed solid ornament, the partially hollow ornament communicating with the vesicle interior. Their description is, however, taxonomically informal and should not be treated as an emendation to the diagnosis. The proposal was not justified by any fur-

ther morphological observations, and the only example of species transferred to the genus (from *Micrhystridium*) was *Buedingiisphaeridium triassicum* (Jansonius, 1962) comb.nov. (Staplin *et al.*, 1965). The species, as illustrated by Jansonius (1962, p. 85, Pl. 16:57) does not show any hollow ornamentation elements, but solid granulae or projections on the surface of the vesicle. According to the emendation of *Buedingiisphaeridium* by Sarjeant & Stancliffe (1994, pp. 24, 25, 66), the genus *Adara* should be rejected and its species transferred to *Buedingiisphaeridium* and *Polygonium*. This reassessment is inconsistent with the taxonomic concepts that are followed herein. *Buedingiisphaeridium* Schaarschmidt, 1963, is considered to be a junior synonym of *Lophosphaeridium* Timofeev, 1959, *ex* Downie, 1963, emend. Lister, 1970. The genus *Adara* Fombella, 1977, emend. Martin, 1981, is maintained as a taxonomically valid and distinctive genus. It comprises the following species: *Adara matutina* Fombella, 1977 (type species), *A. alea* Martin, 1981, *A. longispinosa* Fatka, 1989, and *A. undulata* n.sp. *A. denticulata* Tongiorgi, 1988, *in* Bagnoli *et al.* 1988 is a junior synonym of *A. alea* (Albani *et al.*, 1991).

Palaeontological descriptions

Suprageneric, subgeneric and subspecific taxonomic categories are not recognized among the microfossils described here. In the following chapter, form-genera and form-species are arranged in alphabetic order reflecting their informal taxonomic status (Evitt 1963; Downie *et al.* 1963; Loeblich 1970). Some acritarch genera (e.g., *Leiosphaeridia*, *Tasmanites*, *Cymatiosphaera*, *Pterospermella*, *Dictyotidium*) are regarded as phycomata of prasinophycean algae (Division Prasinophyta, Family Prasinophyceae), or more generally, as being related to Recent Prasinophyta (Jux 1971; Loeblich & Wicander 1976; Tappan 1980; Martin 1993; Colbath & Grenfell 1995). Their possible affinity to modern green algae is not questioned, but for convenience these microfossils are described among other acritarch taxa of unknown systematic position.

Formal acritarch zones are established for the Lower Cambrian and are correlated with the faunal zones (Moczydłowska 1991). In the Middle and Upper Cambrian, several formal and informal acritarch zones have been proposed (Vanguestaine 1974, 1978, 1992; Martin & Dean 1981, 1984, 1988; Volkova 1990), however, they are not always directly related to the faunal zones or are regional and not correlated between each other. For these reasons they are not identified here.

The recognition of biozones in the Upper Silesian successions is based on diagnostic acritarch assemblages. These are correlated directly with trilobite zones or superzones elsewhere and so have good independent control on their age. The exception is the Goczałkowice succession, where olenellid trilobites of the Lower Cambrian *Holmia kjerulfi* Zone have been recorded (Orłowski 1975), providing direct evidence for correlation. Therefore the attribution of the strata to the formal trilobite zones under the entry 'Present record' should be understood to indicate time-equivalence to these zones (because of lack of the trilobite index taxa).The expression 'time-equivalent to ...' is omitted to avoid a multiple repetition, but it should be read as such. The zonal name *Acadoparadoxides oelandicus* replaces the name of the trilobite zone *Eccaparadoxides oelandicus,* and the *Acadoparadoxides pinus* Zone replaces the former *Eccaparadoxides pinus* Zone (Young *et al.* 1994) under the entry 'Occurrence and stratigraphic range'.

The type specimens and illustrated microfossil specimens are housed in the collections of *Paleontologiska Museet* at Uppsala University, carrying prefix PMU-PL. followed by repository number and microscopic slide coordinates. The position of the specimens is given by England Finder coordinates. All photomicrographs were taken using interference contrast, under oil immersion, on a transmitted light Leitz Wetzlar Dialux 20 microscope.

Genus *Actinotodissus* Loeblich & Tappan, 1978

Type species. – *Actinotodissus longitaliosus* Loeblich & Tappan, 1978, pp. 1236–1238, 1241–1242, Pl. 2:1–5; USA, Oklahoma, Murray County, Bromide Formation, Mountain Lake Member, Middle Ordovician, Llandeilo (Loeblich & Tappan 1978).

Actinotodissus? sp.

Fig. 20A

Material. – One poorly preserved specimen.

Description. – Vesicle oval in outline with processes located in the polar portions. The processes are long, robust, tubular, tapering distally, simple, and occasionally divided at the tips. The interior cavity of the processes is connected with the vesicle cavity.

Dimensions. – Diameter of vesicle 14×23 µm, length of processes 11–14 µm.

Present record. – The Upper Silesia area, Sosnowiec IG-1 borehole, Sosnowiec formation at a depth of 3174.0–3174.7 m, lower Upper Cambrian.

Genus *Adara* Fombella, 1977, emended Martin, 1981

Type species. – *Adara matutina* Fombella, 1977, p. 117, Pl. 1:16; Spain, the Cantabrian Mountains, Province of León, the Oville Formation, lower Middle Cambrian (Fombella 1977, Martin *in* Martin & Dean 1981).

Remarks. – The taxonomic status of the type species of the genus was recently confused by Sarjeant & Stancliffe (1994, p. 25). These authors transferred *Adara matutina* to *Buedingiisphaeridium,* referring to it as *Buedingiisphaeridium matutinum* (Fombella 1977, p. 117, pl. 1, Fig. 16) comb.nov., and placing it among the 'accepted species' of their newly emended genus. However, they also transferred it to *Polygonium* (Sarjeant & Stancliffe 1994, p. 44). Apart from the inconsistency, both transfers were based on a wrong assumption, not reinforced by examination, that the processes of *A. matutina* Fombella, 1977, might be of variable nature, i.e. solid, hollow or partially hollow. The latter feature was included in the newly proposed emendation of *Buedingiisphaeridium* Schaarschmidt, 1963, by Sarjeant & Stancliffe (1994, p. 24). The processes in *A. matutina* Fombella, 1977 are exclusively hollow, however, and freely communicate with the vesicle cavity, a feature that is diagnostic for the genus *Adara* Fombella, 1977, emend. Martin, 1981, and contrasts with those in *Buedingiisphaeridium* Schaarschmidt, 1963, which are massive.

Adara alea Martin, 1981

Fig. 21A–C

Synonymy. – □1981 *Adara alea* sp. nov – Martin, *in* Martin & Dean, p. 16, Pls. 1:20–22; 4:7, 9, 10. □1981 *Celtiberium* cf. *geminum* Fombella – Erkmen & Bozdoğan, pp. 49–50, Pl. 1:4. □1981 *Celtiberium geminum* Fombella – Erkmen & Bozdoğan, pp. 49–50, Pl. 1:22. □1984 *Adara alea* Martin *in* Martin & Dean, 1981 – Martin *in* Martin & Dean, Pl. 57:3; Figs. 3, 11. □1988 *Adara alea* – Martin & Dean, Pl. 11:3, 5, 6, 8. □1988 *Adara denticulata* Tongiorgi sp.nov. – Bagnoli *et al.*, pp. 183–184, Pls. 25:1–5; 26:5, 6, 7. □1991 *Adara alea* Martin 1981 – Albani *et al.*, p. 265, Pl. 1:1. □1994 *Polygonium aleum* (Martin *in* Martin & Dean 1981, p. 16, Pls. 1:20–21; 4:7, 9–10) comb.nov. – Sarjeant & Stancliffe, p. 43. □1994 *Adara alea* Martin *in* Martin & Dean, 1981 – Martin *in* Young *et al.*, Fig. 11m, n, p.

Material. – Eighty-nine well-preserved specimens.

Description. – Vesicles circular to oval in outline, originally spherical, with a smooth wall and numerous, robust, short and conical truncated processes. The bases of the processes may possess radially arranged wrinkles because

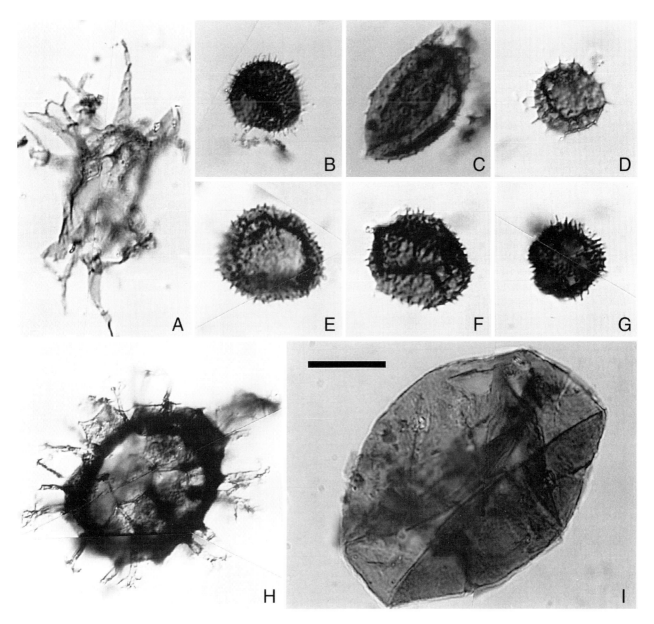

Fig. 20. □A. *Actinotodissus* sp. PMU-Pl.1-Q/48/3, showing different terminations of the polar processes, simple and a few dichotomizing ones. □B. *Asteridium lanatum* (Volkova) Moczydłowska. PMU-Pl.2-U/31/2, displaying diagnostic dense cover of hair-like processes. □C. *Asteridium tornatum* (Volkova) Moczydłowska. PMU-Pl.3-K/26. □D. *Asteridium spinosum* (Volkova) Moczydłowska. PMU-Pl.4-P/34/2. □E–F. *Asteridium solidum* n.sp. E, PMU-Pl.5-L/47/1, holotype, showing solid and robust processes with widened bases and sharp tips; F, PMU-Pl.6-U/37. □G. *Asteridium pilare* n.sp. PMU-Pl.7-W/36/1, holotype; displaying solid and stiff pillar-shaped processes with capitate tips. □H. *Aryballomorpha* sp. PMU-Pl.8-I/45; showing characteristic wide neck-shaped tubular extension. The bases of processes are constricted. □I. *Leiosphaeridia* sp. PMU-Pl.9-T/27/3. A–H from the Sosnowiec IG-1 borehole, Sosnowiec formation; A, F, G, depth 3174.0–3174.7 m, and H, depth 3166.3–3167.7 m, Upper Cambrian; B, C, depth 3403.5–3407.2 m, and D, E. depth 3365.2–3372.3 m, Middle Cambrian, *Acadoparadoxides oelandicus* Superzone; I, Goczałkowice IG-1 borehole, depth 2766.8–2771.0 m, Goczałkowice formation, Middle Cambrian, *Acadoparadoxides oelandicus* Superzone. Scale bar in I equals 12 μm for A–G; 20 μm for H; 14 μm for I.

of compression. The distal ends of the processes are rounded or blunt. The processes are hollow and freely connected with the inner cavity of the vesicle. Occasionally, a translucent membrane stretches between processes.

Dimensions. – *N*=26. Vesicle diameter 23–40 μm, length of processes 2–5 μm.

Present record. – The Upper Silesia area, Sosnowiec IG-1 borehole, Sosnowiec formation at depths of 3210.0–3212.0 m and 3203.5–3205.5 m, Middle Cambrian, *Paradoxides paradoxissimus* and *P. forchhammeri* Superzones, respectively; 3174.0–3174.7 m, lower Upper Cambrian,

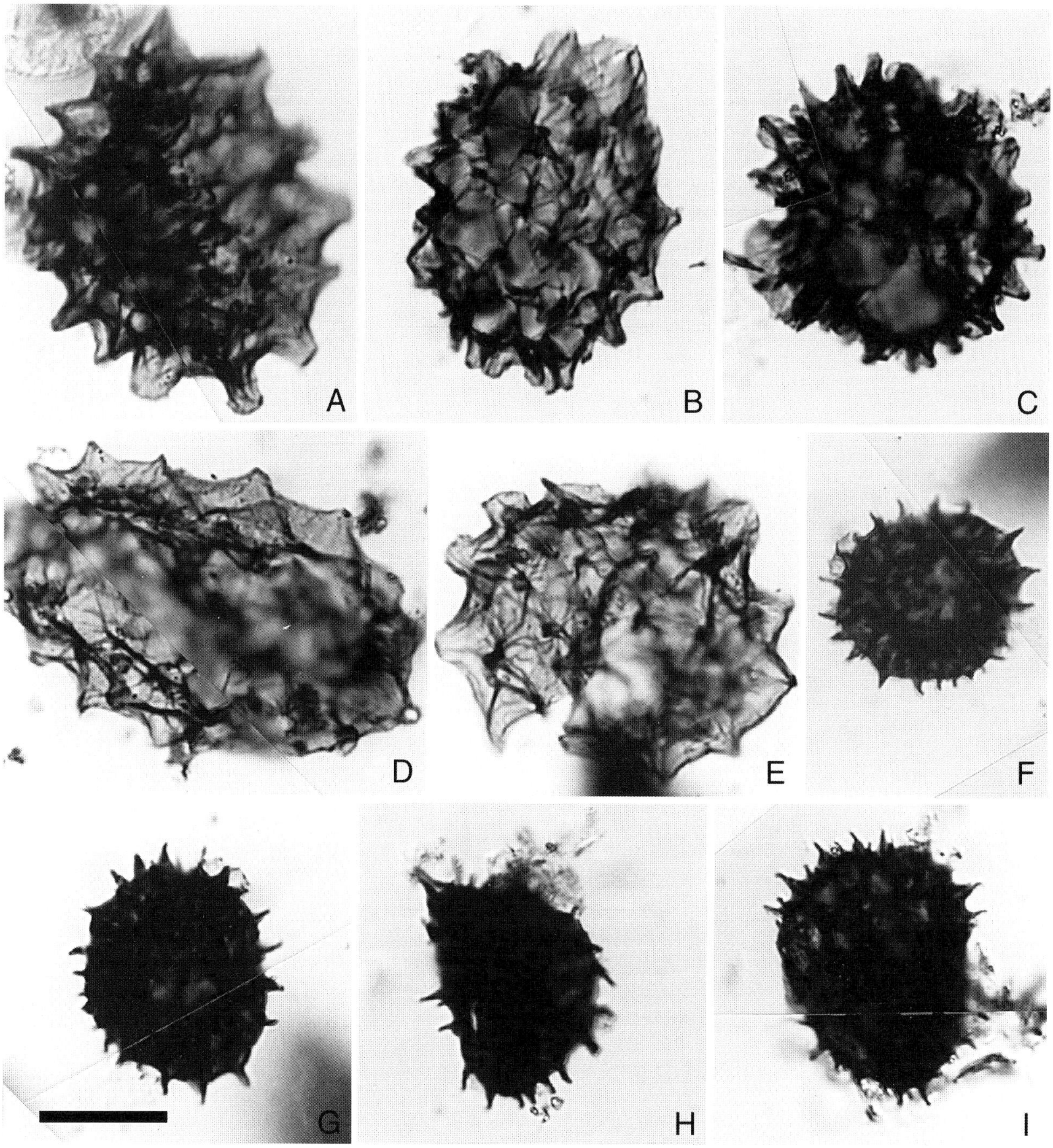

Fig. 21. □A–C. *Adara alea* Martin. A, PMU-Pl.10-X/48/4; B. PMU-Pl.11-K/41/2, showing short conical processes with variable, blunt or rounded, terminations; C, PMU-Pl.12-T/33, a translucent membrane stretched between processes is preserved in lower left quadrant of specimen. □D–E. *Adara undulata* n.sp. D, PMU-Pl.13-P/28/4, holotype; E, PMU-Pl.14-K/31; showing vesicles with hummocky outline formed by short and hollow protrusions with very wide bases. □F–I. *Celtiberium? papillatum* n.sp. F, PMU-Pl.15-P/31/3, holotype, showing evenly distributed processes with free connection into the central body; G, PMU-Pl. 16-H/34; H, PMU-Pl.17-W/47; I, PMU-Pl.18-Y/27/2; exhibiting heteromorphic processes, regular conical in shape and papilliform with swollen bases. All specimens from the Sosnowiec IG-1 borehole, depth 3174.0–3174.7 m, Sosnowiec formation, lower Upper Cambrian. Scale bar in G equals 12 µm for all micrographs.

Occurrence and stratigraphic range. – Canada, Newfoundland, Random Island and Manuels River, Manuels River Formation, Middle Cambrian, *Paradoxides paradoxissimus* Zone (uppermost *Tomagnostus fissus* and *Ptychag-* *nostus atavus* Zone to upper *Ptychagnostus punctuosus* Zone) (Martin & Dean 1981, 1984, 1988). SE Turkey, Mardin–Derik area, Sosink Formation, Middle Cambrian (Erkmen & Bozdoğan 1981), time equivalent to *P. para-*

doxissimus Superzone. Sweden, Öland Island, Furuhäll section at Borgholm, Djupvik Formation, Middle Cambrian, *Paradoxides paradoxissimus* Superzone (Bagnoli *et al.* 1988). Tunisia, Tt 1 borehole, Sidi–Toui Formation, Middle Cambrian, equivalent to upper *Paradoxides paradoxissimus* Superzone (Albani *et al.* 1991). NW Wales, St. Tudwal's Peninsula and St. Tudwal's Island East, Nant-y-big Formation, Middle Cambrian, *Paradoxides paradoxissimus* Superzone (Young *et al.*, 1994).

Adara undulata n.sp.

Fig. 21D–E

Holotype. – Specimen PMU-Pl.13-P/28/4; Fig. 21D.

Synonymy. – □*nomen nudum* 1979 *Celtiberium ondulatum* n.sp. – Fombella, Pl. 2:23.

Derivation of name. – From Latin *undulatus* – wavy; referring to the wavy outline of the vesicle, formed by low protrusions.

Locus typicus. – The Upper Silesia area, Sosnowiec IG-1 borehole (Figs. 2 and 3).

Stratum typicum. – Alternating mudstones and fine-grained sandstones of the Sosnowiec formation at a depth of 3174.0–3174.7 m, lower Upper Cambrian.

Material. – Thirty-two well-preserved specimens.

Diagnosis. – Vesicles circular to oval in outline, originally spherical, having a smooth wall and numerous rigid and short conical processes or protrusions of the vesicle wall, forming a hummocky outline. The processes (protrusions) have wide bases located closely to each other. They possess narrow, rounded or blunt tips and are hollow and connected with the cavity of the vesicle.

Dimensions. – $N=20$. Vesicle diameter 20–45 μm, length of processes 2–4 μm.

Remarks. – A senior but invalidly published synonym of *A. undulata* n.sp. was recognized as *Celtiberium ondulatum* Fombella 1979 (Fombella 1979, p. 4), a *nomen nudum* never formally described. The diagnostic features of the genus *Celtiberium* Fombella, 1977 (Fombella 1977, p. 117; 1978, pp. 250–251) include the presence of long processes with rounded terminations, whose length was estimated to 20% of the diameter of the central body of the vesicle. Contrary to this, *C.* 'ondulatum' has very low conical processes which correspond to the description of the genus *Adara* Fombella, 1977, emend. Martin, 1981. Accordingly, *C.* 'ondulatum' should be transferred to the genus *Adara*.

The new described species differs from *A. alea* Martin, 1981, by having processes whose bases are wider and form hummocky protrusions on the wall.

Present record. – The Upper Silesia area, Sosnowiec IG-1 borehole, Sosnowiec formation at depths of 3210.0–3212.0 m and 3174.0–3174.7 m, Middle Cambrian, *Paradoxides paradoxissimus* Superzone and lower Upper Cambrian, respectively.

Occurrence and stratigraphic range. – Spain, Cantabrian Mountains at Sebares, the Oville Formation referred to as Middle Cambrian – Lower Tremadocian by Fombella (1979). The stratigraphic range of the formation at this locality is here considered to be only Middle Cambrian, probably equivalent to *P. paradoxissimus* Superzone (see under 'Stratigraphic ranges').

Genus *Alliumella* Vanderflit, 1971

Type species. – *Alliumella baltica* Vanderflit, 1971, *in* Umnova & Vanderflit, pp. 69–70, Pl. 2:42–48; Russia, Kaliningrad area, Pirita Formation, Lower Cambrian.

Alliumella baltica Vanderflit, 1971

Material. – Six poorly preserved specimens.

Synonymy and description. – See Moczydłowska 1991, p. 46.

Dimensions. – $N=6$. Vesicle diameter 14–18 μm, process length 7–9 μm.

Present record. – The Upper Silesia area, Sosnowiec IG-1 borehole, the Sosnowiec formation at a depth of 3365.2–3372.3 m, Middle Cambrian, *Acadoparadoxides oelandicus* Superzone.

Occurrence and stratigraphic range. – Belarus, Latvia, the Ukraine and Russia, Lower Cambrian, Talsy, Vergale and Rausve horizons; Latvia, Middle Cambrian, Kibartai horizon (Volkova *et al.* 1979). Kazakhstan, Maly Karatau, Shabakty Formation, Lower Cambrian (Ogurtsova 1985). Sweden, Västergötland, File Haidar Formation, Lingulid Sandstone Member, Lower Cambrian, *Holmia kjerulfi* and *Proampyx linnarssoni* Zones (Moczydłowska & Vidal 1986; Moczydłowska 1991). Belgium, Rocroi Massif and Stavelot Massif, Middle Cambrian (Vanguestaine 1974, 1978). Poland, East European Platform, Lublin Slope, Kaplonosy and Radzyń formations, and Podlasie Depression, 'Holmia and Protolenus' zones, Lower Cambrian, *Holmia kjerulfi* and *Protolenus* Zones; Lublin Slope, Kostrzyń Formation, Middle Cambrian, *Acadoparadoxides oelandicus* Zone (Volkova *et al.* 1979; Moczydłowska 1991).

Genus *Aryballomorpha* Martin & Yin Leiming, 1988

Type species. – Aryballomorpha grootaertii (Martin, 1984) Martin & Yin Leiming, 1988 (Martin & Yin Leiming 1988, pp. 113–114; holotype in Martin 1984, Pl. 58.1, Figs. 8–9); Canada, Alberta, the Rocky Mountains at Wilcox Pass, the Survey Peak Formation, Upper Tremadoc.

Aryballomorpha sp.

Fig. 20H

Material. – A single, fairly well-preserved specimen.

Description. – Vesicle originally spherical, ovoid in outline, having distinct tubular extension in shape of neck and bearing evenly distributed processes. The vesicle wall is single-layered, and the vesicle extension is terminated by the opening. The processes are cylindrical and ramifying at the distal portions up to the third order of small pinnulae. The processes are hollow, but communication with the vesicle cavity has not been observed. Some process bases are plugged.

Dimensions. – N=1. Vesicle diameter 36×48 μm, length of processes 10–15 μm, width of the neck-like extension 10 μm.

Remarks. – It is difficult to determine if the process cavities are in contact with the vesicle interior. The vesicle wall is more dense and darker than the processes, and it appears to be forming a plug between the vesicle and processes. However, the plug might be secondary, a result of fossilization, as observed in *Vogtlandia* sp., *Stelliferidium robustum* n.sp., *Multiplicisphaeridium sosnowiecense* n.sp., and *M. varietatis* n.sp. Well-preserved specimens of these species have free communication between process cavities and the vesicle, whereas plugs are observed in poorly preserved specimens.

The recorded specimen differs from *Aryballomorpha grootaertii* (Martin, 1984) Martin & Yin Leiming, 1988, by having fewer processes and by lacking anastomosing tips of the processes to form a peripheral meshwork.

Present record. – The Upper Silesia area, Sosnowiec IG-1 borehole, Sosnowiec formation at a depth of 3166.3–3167.7 m, lower Upper Cambrian.

Genus *Asteridium* Moczydłowska, 1991

Type species. – Asteridium lanatum (Volkova, 1969) Moczydłowska, 1991, p. 47 (=*Micrhystridium lanatum* Volkova, 1969b, p. 227, Pl. 50:27–28); Poland, East European Platform, Radzyń formation, Lower Cambrian, *Holmia kjerulfi* Zone.

Asteridium lanatum (Volkova, 1969) Moczydłowska, 1991

Fig. 20B

Material. – Fifteen well-preserved specimens.

Synonymy and description. – See Moczydłowska 1991, p. 47. Additionally: □*non* 1986 *Micrhystridium lanatum* Volkova 1969 – Welsch, Pl. 1:15–17. □1989 *Micrhystridium lanatum* Volkova, 1969 – Hagenfeldt 1989a, pp. 74–77, Pl. 3:12. □1994 *Filisphaeridium lanatum* (Volkova 1969, p. 227, Pl. 50:27, 28) comb. nov – Sarjeant & Stancliffe, p. 21, 30. □1995 *Asteridium lanatum* (Volkova 1969) Moczydłowska 1991 – Vidal *et al.*, Fig. 5:9.

Dimensions. – N=5. Vesicle diameter 7–14 μm, process length 2–3 μm.

Present record. – The Upper Silesia area, Sosnowiec IG-1 borehole, Sosnowiec formation at a depth of 3403.5–3407.2 m, Middle Cambrian, *Acadoparadoxides oelandicus* Superzone.

Occurrence and stratigraphic range. – The species has been recorded worldwide throughout the Lower Cambrian (see comprehensive compilation by Moczydłowska 1991) and in the lowermost Middle Cambrian. Additional occurrences are: Sweden, Island of Gotland, Grötlingbo-1 borehole, and Gotska Sandön Island, Gotska Sandön borehole, File Haidar Formation; the Gulf of Bothnia, Finngrundet borehole, Söderfjärden Formation; Finland, the Vaasa area, Söderfjärden-3 borehole, Söderfjärden Formation; Lower Cambrian, *Holmia kjerulfi* and *Protolenus Zones*, and Middle Cambrian, *Eccaparadoxides insularis* Zone (Hagenfeldt 1989a). North Greenland, Peary Land, Buen Formation, Lower Cambrian, *Holmia kjerulfi* and *Protolenus* Zones (Vidal & Peel 1993). Eastern Siberia, Yakutia, the Kharaulakh Mountains at Chekurovka, the Tyusersk Formation, Lower Cambrian, Tommotian Stage, *Dokidocyathus regularis* Zone (Vidal *et al.* 1995).

Asteridium pilare n.sp.

Fig. 20G

Holotype. – Specimen PMU-Pl.7-W/36/1; Fig. 20G.

Derivation of name. – From Latin *pila* – column, pilaster; *pilaris* – columnar; referring to the shape of the processes.

Locus typicus. – The Upper Silesia area, Sosnowiec IG-1 borehole (Figs. 2 and 3).

Stratum typicum. – Alternating mudstones and fine-grained sandstones in the Sosnowiec formation at a depth of 3174.0–3174.7 m, Upper Cambrian.

Material. – Three satisfactorily preserved specimens.

Diagnosis. – Vesicles circular in outline, originally spherical, bearing numerous, evenly distributed uniform processes. The processes are solid, short, stiff in appearance and pillar-shaped with capitate and occasionally rounded tips.

Dimensions. – *N*=3. Vesicle diameter 14–23 μm (holotype 14 μm), process length around 2 μm.

Remarks. – The shape of the processes is very similar to that of *Acrum*? *araxisii* Di Milia 1991 (p. 133). The major difference at the generic level is the presence of a membrane between the processes in *A.*? *araxisii*.

Present record. – As for the holotype.

Asteridium solidum n.sp.

Fig. 20E–F

Holotype. – Specimen PMU-Pl.5-L/47/1; Fig. 20E.

Synonymy. – 1990 *Micrhystridium* sp. – Volkova, Pl. 3:8.

Derivation of name. – From Latin adj. *solidus* – massive, referring to the general appearance of the processes.

Locus typicus. – The Upper Silesia area, Sosnowiec IG-1 borehole (Figs. 2 and 3).

Stratum typicum. – Alternating mudstones and fine-grained sandstones in the Sosnowiec formation at a depth of 3365.2–3372.3 m; Middle Cambrian, *Acadoparadoxides oelandicus* Superzone.

Material. – Four fairly well-preserved specimens.

Diagnosis. – Vesicles circular to oval in outline, originally spherical, bearing abundant short and robust spines, widened proximally and with sharp tips. The spines are solid and are densely and evenly distributed.

Dimensions. – *N*=4. Vesicle diameter 15–19 μm (holotype 18 μm), process length around 2 μm.

Remarks. – The species differs from *Lophosphaeridium bacilliferum* Vanguestaine, 1974, by having a smaller vesicle diameter and thorn-shaped processes.

Present record. – The Upper Silesia area, Sosnowiec IG-1 borehole, Sosnowiec formation at depths of 3365.2–3372.3 m and 3210.0–3212.0 m, Middle Cambrian *Acadoparadoxides oelandicus* and *Paradoxides paradoxissimus* Superzones, respectively; a depth of 3174.0–3174.7 m, lower Upper Cambrian.

Occurrence and stratigraphic range. – Russia, Moscow Syneclise, Jaroslav area, borehole Tolbukhino-1, Molozhsk Formation, Middle Cambrian, upper *Paradoxides paradoxissimus* to lower *Paradoxides forchhammeri* Superzones (Volkova 1990).

Asteridium spinosum (Volkova, 1969) Moczydłowska, 1991

Fig. 20D

Material. – Forty-seven well-preserved specimens.

Synonymy and description. – See Moczydłowska 1991, p. 48. Additionally: □1989 *Micrhystridium spinosum* Volkova, 1969 – Hagenfeldt 1989a, Pl. 4:3. □1994 *Comasphaeridium spinosum* (Volkova 1969b, p. 229, Pl. 50:14–16) comb.nov. – Sarjeant & Stancliffe, p. 27.

Dimensions. – *N*=10. Vesicle diameter 7–14 μm, length of processes 2–3 μm.

Present record. – The Upper Silesia area, Sosnowiec IG-1 borehole, Sosnowiec formation at depths of 3403.5–3407.2 m, 3365.2–3372.3 m and 3353.3–3359.3 m; the GoczałkowiceIG-1 borehole, Goczałkowice formation at a depth of 2766.8–2771.1 m – Middle Cambrian, *Acadoparadoxides oelandicus* Superzone.

Occurrence and stratigraphic range. – Poland, the East European Platform, Lublin Slope, the Kaplonosy and Radzyń formations and Podlasie Depression, 'Holmia and Protolenus' zones, Lower Cambrian, *Holmia kjerulfi* and *Protolenus* Zones; Lublin Slope, Kostrzyń Formation, Middle Cambrian, *Acadoparadoxides oelandicus* Zone (Volkova 1969a, b; Volkova *et al.* 1979). Latvia and the Ukraine, Lower Cambrian, Vergale and Rausve horizons and Latvia, Middle Cambrian, Kibartai horizon (Volkova *et al.* 1979). Kazakhstan, Maly Karatau, Shabakty Formation, Lower Cambrian (Ogurtsova 1985). Sweden, Kalmarsund region, '*Mobergella* Sandstone', Lower Cambrian, *Holmia kjerulfi* Zone (Moczydłowska & Vidal 1986; Moczydłowska 1991). Scotland, Fucoid Beds, Lower Cambrian (Downie 1982). Ireland, Bray Group, Thulla Fromation, Lower–Middle Cambrian (Gardiner & Vanguestaine 1971). East Greenland, Ella Ø, Bastion Formation, Lower Cambrian (Downie 1982). Sweden, Gotland Island, Grötlingbo-1 borehole and Gotska Sandön Island, File Haidar Formation and 'oelandicus beds', Lower Cambrian, *Holmia kjerulfi* and *Proampyx linnarsoni* Zones and Middle Cambrian, *Acadoparadoxides oelandicus* Superzone; Gulf of Bothnia, Finngrundet borehole, Söderfjärden Formation, Lower Cambrian, *Proampyx linnarsoni* Zone and Middle Cambrian, *Acadoparadoxides oelandicus* Superzone; Närke area at Kvarntorp, 'oelandicus beds', Middle Cambrian, *Acadopara-*

doxides oelandicus Superzone (Hagenfeldt 1989a). Finland, Vaasa area, Söderfjärden-3 borehole, Söderfjärden Formation, Lower Cambrian, *Holmia kjerulfi* and *Proampyx linnarsoni* Zones (Hagenfeldt 1989a).

Asteridium tornatum (Volkova, 1968) Moczydłowska, 1991

Fig. 20C

Material. – Sixteen poorly preserved specimens.

Synonymy and description. – See Moczydłowska 1991, pp. 48–49. Additional: □*non* 1986 *Micrhystridium tornatum* Volkova 1968 – Welsch, pp. 60–61, Pl. 2:5–8. □1989 *Micrhystridium tornatum* Volkova, 1968 – Hagenfeldt 1989a, pp. 85, 92–93, 95–97, Pl. 4:4. □1994 *Filisphaeridium tornatum* (Volkova 1968, p. 21, Pls. 4:1–4; 10:8) comb.nov. – Sarjeant & Stancliffe, pp. 22 and 31.

Dimensions. – $N = 10$. Vesicle diameter 14–22 µm, length of processes less than 1 µm.

Present record. – The Upper Silesia area, Sosnowiec IG-1 borehole, Sosnowiec formation at depths of 3403.5–3407.2 m and 3365.2–3372.3 m; Middle Cambrian, *Acadoparadoxides oelandicus* Superzone.

Occurrence and stratigraphic range. – The species has been recorded having a worldwide distribution from the base of the Lower Cambrian to the Middle Cambrian, *Acadoparadoxides oelandicus* Superzone (for compilation see Moczydłowska 1991, p. 49). New occurrences are in Sweden, Island of Gotland, Grötlingbo-1 borehole, and Gotska Sandön Island, Gotska Sandön borehole, File Haidar Formation and the 'oelandicus beds'; the Gulf of Bothnia, Finngrundet borehole, Söderfjärden Formation and 'oelandicus beds'; Lower Cambrian, *Holmia kjerulfi* and *Protolenus Zones*, and Middle Cambrian, *Acadoparadoxides oelandicus* Superzone, respectively (Hagenfeldt 1989a). North Greenland, Peary Land, Buen Formation, Lower Cambrian, *Holmia kjerulfi* and *Protolenus* Zones (Vidal & Peel 1993). Eastern Siberia, Yakutia, the Kharaulakh Mountains at Chekurovka, the Tyusersk Formation, Lower Cambrian, Tommotian Stage, *Dokidocyathus regularis* Zone (Vidal *et al.* 1995).

Asteridium sp.

Material. – More than 50 specimens in variable states of preservation.

Description. – Vesicles circular to oval in outline, originally spherical, bearing thin but rigid and solid processes with sharp tips.

Dimensions. – Vesicle diameter ranges 7–16 µm, process length 5–7 µm.

Present record. – The Upper Silesia area, Sosnowiec IG-1 borehole, Sosnowiec formation at various stratigraphic levels within the interval of 3174.0–3407.2 m, Middle–Upper Cambrian; Goczałkowice IG-1 borehole, Goczałkowice formation, at a depth of 2766.8–2771.1 m, Middle Cambrian, *Acadoparadoxides oelandicus* Superzone.

Genus *Celtiberium* Fombella, 1977

Type species. – *Celtiberium geminum* Fombella, 1977, pp. 117–118, Pl. 1:10, 11; Fig. 1:9; Spain, Cantabrian Mountains, Province of León, Oville Formation, lower Middle Cambrian.

Remarks. – The genus *Celtiberium* (Fombella 1977, p. 117, 1978, pp. 250–251) differs from *Polygonium* Vavrdová, 1966, by having a circular to subcircular outline of the vesicle, contrary to the polygonal outline in the latter genus, and by having narrower, less distinctive proximal portions of the processes. These clear morphological differences are sufficient to regard *Celtiberium* a distinct taxon, not a junior synonym of *Polygonium* as assumed by Sarjeant & Stancliffe (1994, pp. 42 and 67). Accordingly, their proposed new combinations of *Celtiberium dedalinum* Fombella, 1978, *C. clarum* Fombella, 1978, and *C. geminum* Fombella, 1977, as species of *Polygonium* are not recognized.

Celtiberium? papillatum n.sp.

Fig. 21F–I

Holotype. – Specimen PMU-Pl.15-P/31/1; Fig. 21F.

Derivation of name. – From Latin *papilla* – nipple, in reference to the papilliform shape of the processes.

Locus typicus. – The Upper Silesia area, Sosnowiec IG-1 borehole (Figs. 2 and 3).

Stratum typicum. – Alternating mudstones and fine-grained sandstones in the Sosnowiec formation at a depth of 3174.0–3174.7 m, Upper Cambrian.

Material. – Twenty-one well-preserved and satisfactorily preserved specimens.

Diagnosis. – Vesicles circular to subcircular in outline, originally sphaerical, having numerous, evenly distributed, short heteromorphic processes. Some processes are regularly conical in shape, and others are papilliform with swollen (=rounded) bases and tapering terminations; the papilliform processes are also fewer. All processes are hollow and connected with the inner cavity of the vesicle.

Dimensions. – $N=12$. Diameter of the vesicle 15–27 μm (holotype 19 μm), length of the processes 2–4 μm (holotype 2 μm).

Remarks. – *Celtiberium*? *papillatum* n.sp. differs from other species of *Celtiberium* by having tapering distal portions of the processes; it is therefore tentatively attributed to the genus. It differs from *C. clarum* Fombella, 1978, by having papilliform proceses and from *C. dedalinum* Fombella, 1978, in having wider process bases.

Present record. – As for the holotype.

Genus *Comasphaeridium* Staplin *et al.*, 1965

Type species. – *Comasphaeridium cometes* (Valensi, 1949) Staplin *et al.* 1965, p. 192 (=*Micrhystridium cometes* Valensi, 1949, p. 545, Fig. 5:6); France, Middle Jurassic.

Remarks. – The emendation of *Comasphaeridium* proposed by Sarjeant & Stancliffe (1994, p. 25) is not followed here, because no substantial features has been added, the emendation being confined to an arbitrary chosen limit of the number of processes and the ratio of the length of processes and vesicle diameter. These limitations are superfluous and taxonomically meaningless (see under 'Taxonomic comments').

Comasphaeridium gogense (Downie, 1982) Sarjeant & Stancliffe, 1994

Fig. 31D

Synonymy. – □*nomen nudum* 1979 *Comasphaeridium filiforme* n.sp. – Fombella, Pl. 3:47. □1982 *Micrhystridium gogensis* sp.nov. – Downie, p. 261, Figs. 4, 6o. □*nomen nudum* 1987 *Comasphaeridium filiforme* Fombella, 1979 – Fombella, Pl. 1:10. □1982? *Gorgonisphaeridium* sp. F – Wood & Clendening, p. 262, Pl. 2:10. □1994 *Comasphaeridium gogense* (Downie 1982, p. 261, Figs. 4, 6o) comb.nov. – Sarjeant & Stancliffe, p. 26.

Material. – Seven well-preserved specimens.

Description. – Minute vesicles circular to oval in outline, originally spherical, bearing abundant and closely arranged homomorphic solid processes. The processes are of equal length, filiform with slightly widened bases and truncated tips.

Dimensions. – $N=7$. Vesicle diameter 7–9 μm, length of processes 2–3 μm.

Remarks. – The nature of the processes (solid or hollow) was not determined in the diagnosis of *Micrhystridium*

gogensis (Downie 1982). Its attribution to the genus *Micrhystridium* did not unequivocally establish the morphology of the processes. This is because several morphotypes were previously included in this genus (Moczydłowska 1991). The processes of the holotype illustrated by Downie (1982, Fig. 6o) are narrow and seem to be solid. Microfossils described by Wood & Clendening (1982) that are regarded here as conspecific have solid processes. Under the light microscope, specimens in the present collection also appear to be solid.

Present record. – The Upper Silesia area, Sosnowiec IG-1 borehole, Sosnowiec formation at depths of 3403.5–3407.2 m and 3365.2–3372.3 m; Middle Cambrian, *Acadoparadoxides oelandicus* Superzone.

Occurrence and stratigraphic range. – Canada, Alberta, Mount Eisenhower, Gog Formation, Lower Cambrian, equivalent to *Holmia kjerulfi* and *Proampyx linnarssoni* Zones (Downie 1982). Spain, Cantabrian Mountains, Province of León, locality uncertain (probably Lois), and Vozmediano, the Oville Formation, lower part of Middle Cambrian (Fombella 1979) and alleged Upper Cambrian (Fombella 1987). The extension of the stratigraphic range of the Oville Formation to Upper Cambrian is in doubt and is not included here (see under 'Stratigraphic ranges').

Comasphaeridium longispinosum Hagenfeldt, 1989

Fig. 22D

Synonymy. – □1989 *Comasphaeridium longispinosum* Hagenfeldt n. sp. – Hagenfeldt 1989 b, pp. 192–194, Pl. 1:5–6. □*non* 1993 *Comasphaeridium longispinosum* Vidal n.sp. – Vidal & Peel, pp. 19–21, Fig. 6a. □1994 *Comasphaeridium longispinosum* Hagenfeldt, 1989b – Martin in Young *et al.*, Fig. 11f, l.

Material. – Two poorly preserved specimens.

Description. – Vesicles oval in outline, originally spherical, covered by abundant and solid processes, about equal to or longer than the diameter of vesicle. The processes are filiform, thin and flexible, arising perpendicularly from the vesicle wall.

Dimensions. – $N=2$. Vesicle diameter 18×29 μm and 20×30 μm, length of processes 23–27 μm.

Remarks. – The species *Comasphaeridium longispinosum* Vidal, 1993 (in Vidal & Peel 1993), is a junior homonym of *Comasphaeridium longispinosum* Hagenfeldt, 1989 (Hagenfeldt 1989b), but it is a separate species and should be renamed.

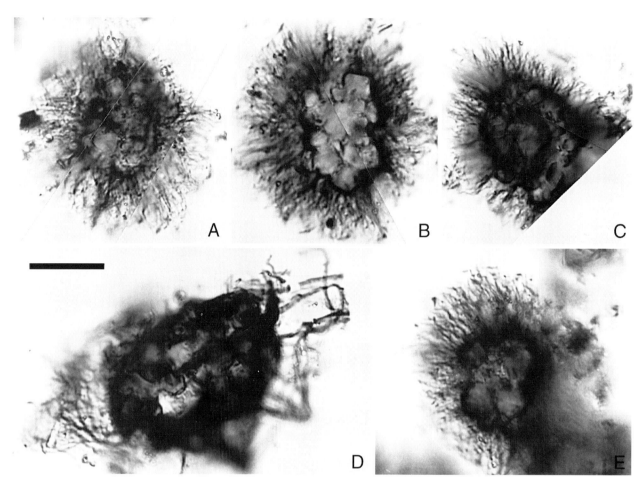

Fig. 22. □A–C, E. *Comasphaeridium silesiense* n.sp. A, PMU-Pl.19-L/38/4, holotype, vesicle densely covered by processes that are slender and flexible; B, PMU-Pl.20-G/46/4, paratype, central body of specimen is distorted by pyrite pseudomorphs; C, PMU-Pl.21-I/46; E, PMU-Pl.22-H/31/1; the specimen shows clearly solid processes of equal length. □D. *Comasphaeridium longispinosum* Hagenfeldt. PMU-Pl.23-I/33/4, showing extremely long flexible processes. All specimens from the Sosnowiec IG-1 borehole, depth 3174.0–3174.7 m, Sosnowiec formation, lower Upper Cambrian. Scale bar in D equals 12 μm for all micrographs.

Present record. – The Upper Silesia area, Sosnowiec IG-1 borehole, Sosnowiec formation at a depth of 3174.0– 3174.7 m; lower Upper Cambrian.

Occurrence and stratigraphic range. – Sweden, south-central area, Närke at Kvarntorp and Gotland Island, Gotska Sandön Island, Gulf of Bothnia, Finnrundet borehole, 'oelandicus beds' and Finland, Vaasa area, Söderfjärden 3 borehole, upper Söderfjärden Formation, Middle Cambrian, *Acadoparadoxides oelandicus* Superzone (Hagenfeldt 1989b). NW Wales, St. Tudwal's Peninsula and St. Tudwal's Island East, Nant-y-big Formation, Middle Cambrian, *Paradoxides paradoxissimus* Superzone (Young *et al.* 1994).

Comasphaeridium silesiense n.sp.

Fig. 22A–C, E

Holotype. – Specimen PMU-Pl.19-L/38/4; Fig. 22A. Paratype specimen PMU-Pl.20-G/46/4; Fig. 22B.

Synonymy. – □*nomen nudum* 1979 *Comasphaeridium vozmedianum* n.sp.; Fombella, p. 4, Pl. 2:29. □1987 *Comasphaeridium strigosum* (Jankauskas), Downie, 1982 – Fombella, Pl. 2:20. □1989 *Comasphaeridium strigosum* (Yankauskas) Downie 1982; Hagenfeldt 1989a, Pl. 2:2. □1991 *Comasphaeridium* sp. – Albani *et al.*, p. 266, Pl. 1:2. □1992 *Comasphaeridium* sp. cf. *C. strigosum* (Jankauskas) Downie, 1982 – Martin, Pl. 6:6, 7. □1993 *Comasphaeridium strigosum* (Yankauskas *in* Yankauskas & Posti) Downie, 1982 – Fombella *et al.* Pl. 3:3. □1995 *Comasphaeridium* n.sp. A – Moczydłowska & Crimes, p. 119, Pl. 1D.

Derivation of name. – Referring to Silesia in southern Poland, type locality of the species.

Locus typicus. – The Upper Silesia area, Sosnowiec IG-1 borehole (Figs. 2 and 3).

Stratum typicum. – Alternating mudstones and sandstones in the Sosnowiec formation at a depth of 3174.0–3174.7 m; Upper Cambrian.

Material. – Sixty-four fairly well-preserved specimens. The central body of many specimens is distorted by imprints of pyrite crystals.

Diagnosis. – Vesicles circular to oval in outline, originally sphaerical, bearing very numerous, evenly distributed and densely arranged, solid, slender and flexible processes. The processes are filose and approximately of equal length.

Dimensions. – $N = 18$. Vesicle diameter 11–25 μm (holotype 17 μm), length of processes 6–13 μm (holotype 9–11 μm).

Remarks. – Acritarchs attributed to this species were left under open nomenclature (*Comasphaeridium* n.sp. A) by Moczydłowska & Crimes (1995). Its stratigraphic range is probably Middle–Upper Cambrian, as estimated from the occurrences of conspecific taxa.

Present record. – The Upper Silesia area, Goczałkowice IG-1 borehole, Goczałkowice formation at a depth of 2766.8–2771.1 m, Middle Cambrian, *Acadoparadoxides oelandicus* Superzone; the Sosnowiec IG-1 borehole, Sosnowiec formation at an interval of 3174.0–3212.0 m, Middle Cambrian, *Paradoxides paradoxissimus* Superzone to lower Upper Cambrian.

Occurrence and stratigraphic range. – Northern Spain, Cantabrian Mountains, Oville Formation referred to as Upper Cambrian (Fombella 1979, 1986, 1987). This stratigraphic range of the Oville Formation is not, however, documented and is probably Middle Cambrian (see under 'Range'). Sweden, Gulf of Bothnia, Finngrundet borehole, Söderfjärden Formation, Middle Cambrian, *Acadoparadoxides oelandicus* Superzone (Hagenfeldt 1989a). Libya, northern Tripolitania, Rhadames Basin, A1-70 borehole, Sidi-Toui Formation, lower Upper Cambrian (Albani *et al.* 1991). Canada, Rocky Mountains, Wilcox Pass, Survey Peak Formation (basal silty member), lowermost Ibexian = transitional Upper Cambrian – Tremadoc or Tremadoc (Martin 1992). Eire, Duncannon area, Co. Wexford, Booley Bay Formation, upper Upper Cambrian (Moczydłowska & Crimes 1995).

Genus *Cristallinium* Vanguestaine, 1978

Type species. – *Cristallinium cambriense* (Slaviková, 1968) Vanguestaine, 1978, pp. 270–271 (=*Dictyotidium cambriense* Slaviková 1968, p. 201, Pl. 2:1 [holotype] and 2:3); Czech Republic, Bohemian Massif, Jince Formation, Middle Cambrian, *Ellipsocephalus hoffi* Subzone.

Remarks. – The generic name *Cristallinium* was considered by Fensome *et al.* (1990, p. 160) to be invalidly published by Vanguestaine (1978, pp. 270–271) since the name of the type species *Cristallinium* (=*Dictyotidium*) *cambriense* (Slavikova, 1968) Vanguenstaine, 1978, was not (in their opinion) validly published. They concluded that the holotype of *D. cambriense* was not designated by Slaviková (1968). In the original designation of the holotype (Slaviková 1968, p. 201), there is a printing error caused by the use of both Roman and Arabic numerals, indicating as holotype the 'specimen figured on plate I, Fig. 11' when it should be 'plate II, Fig. 1'. This minor misprint is easy to identify and enables correct recognition of the holotype from the two micrographs illustrating the species (Slaviková 1968, Pl. 2:1, 3). Hence, Vanguestaine (1978) and subsequent researchers (see synonymy of the species) regarded *Cristallinium* (=*Dictyotidium*) *cambriense* as a validly published species (Martin *in* Young *et al.* 1994; Molyneux & Fensome 1996).

The proposed validation of the genus *Cristallinium* Vanguestaine, 1978, *ex* Fensome *et al.*, 1990, and designation of the new type species, *Cristallinium* (al. *Cymatiosphaera*) *ovillense* (Cramer & Diez, 1972) Fensome *et al.*, 1990, by Fensome *et al.* 1990 (pp. 160–161) is redundant. Additionally, *Cristallinium ovillense* (Cramer & Diez, 1972) had been earlier recognized as a new combination by Martin (1981 *in* Martin & Dean 1981) to replace the invalidly published *Zonosphaeridium ovillensis* (Cramer & Diez 1972). Fensome *et al.* (1990, p. 160) proposed that a new name be substituted for the latter species, *Cristallinium deceptum* sp.nov. At the same time, they recombined *Cymatiosphaera ovillensis* Cramer & Diez, 1972, as *Cristallinium ovillense* (Cramer & Diez, 1972) comb.nov. (Fensome *et al.* 1990, p. 161), though the former species is a junior synonym of *Cristallinium cambriense* (Slaviková, 1968) Vanguestaine, 1978 (and it is retained as such, see below). These taxonomic changes of the genus *Cristallinium* and its species by Fensome *et al.* 1990, following rigorously the recommendations of nomenclatural code, are in practice both confusing and unnecessary. It is reasonable and easier to correct a small misprint than to abandon the type species after more than twenty years of recognition.

Here, the originally recognized genus *Cristallinium* Vanguestaine, 1978, and its species are retained as follows:

C. aciculatum Tongiorgi *in* Bagnoli *et al.* 1988 (pp. 186–188, Pl. 28:3–6).

C. cambriense (Slaviková, 1968) Vanguestaine, 1978; type species (see below).

C. dentatum (Vavrdová, 1976) Li, 1987; Vavrdová 1976, pp. 58–59, Pl. 2:5–6; Li 1987, p. 618, Pl. 68:6.

C. ovillense (Cramer & Diez, 1972) Martin, 1981, *in* Martin & Dean 1981 (see below).

C. pilosum Golub & Volkova, 1985, *in* Volkova & Golub 1985 (pp. 105–106, Pl. 7:10).

C. randomense (Martin, 1981) Martin, 1988, *in* Martin & Dean 1988 (pp. 36–37, Pl. 13:1–17) (see below).

The three species of *Cristallinium*, *C. baculatum*, *C. dubium* and *C. locale*, that were recognized by Volkova (1990) as new taxa are here considered to be conspecific with previously defined species. *C. baculatum* (Volkova 1990, p. 60, Pl. 14:12–13) resembles specimens of *C. randomense* with broken terminal portions of the ornaments, as for instance specimens illustrated by Martin (*in* Martin & Dean 1988, Pl. 13:4, 5, 8), and is considered to be its junior synonym. *C. dubium* (Volkova 1990, p. 61, Pl. 3:1–3) and *C. locale* (Volkova 1990, p. 62, Pl. 3:4, 5, 11) are poorly preserved and do not display convincingly the ascribed diagnostic features.

   C. deceptum Fensome *et al.* 1990 (p. 160), proposed as a taxonomic transfer of *C. ovillense,* is here regarded as redundant.

Cristallinium cambriense (Slaviková, 1968) Vanguestaine, 1978

Fig. 23A–D

Synonymy. – 1968 *Dictyotidium cambriense* n.sp. – Slaviková, p. 201, Pl. 2:1, 3. □1971 *Dictyotidium cambriense* – Gardiner & Vanguestaine, p. 195, Pl. 2:4, 5. □1972 *Cymatiosphaera ovillensis* Cramer & Diez (New Species) – Cramer & Diez, p. 44, Pl. 2:4, 7, 10. □1976 *Cymatiosphaera ovillense* Cramer & Diez – Vavrdová, Pl. 4:8. □1976 *Cymatiosphaera crameri* Slaviková – Vavrdová, Pl. 1:6, 10. □1976 *Cymatiosphaera favosa* Jankauskas, sp. nov – Jankauskas, p. 190, Pl. 25:7, 15. □1976 *Cymatiosphaera lazdynica* Jankauskas, sp. nov – Jankauskas, pp. 190–191, Pl. 25:4, 5, 8, 10. □1976 *Cymatiosphaera cristata* Jankauskas, sp. nov – Jankauskas, p. 191, Pl. 25:18, 21. □1976 *Cymatiosphaera nerisica* Jankauskas, sp. nov – Jankauskas, pp. 191–192, Pl. 25:11, 19. □1977 *Dictyotidium cambriense* Slaviková – Martin, Pl. 4:12. □1978 *Cristallinium cambriense* (Slaviková) Vanguestaine, nov. comb. – Vanguestaine, pp. 270–271, Pls. 2:16, 17; 3:16, 26. □1980 *Dictyotidium* aff. *cambriense* Slaviková – Volkova, Pl., Fig. 7. □1980 *Cymatiosphaera ovillensis* Cramer et Diez – Jankauskas 1980, Pl., Figs. 11, 12. □1985 *Cristallinium cambriense* (Slaviková) Vanguestaine 1978 – Pittau, pp. 180–182, Pls. 7:3; 4:12. □1986 *Cristallinium cambriense* (Slaviková 1968) – Welsch, pp. 75–76, Pl. 3:7–14. □1987 *Cristallinium cambriense* (Slaviková, 1968) – Fombella, Pl. 2:1. □1988 *Cristallinium cambriense* (Slaviková) Vanguestaine 1978 – Bagnoli *et al.*, pp. 188–189, Pls. 25:6–9; 28:1–2. □1988 *Cristallinium cambriense* (Slaviková) Vanguestaine, 1978 – Martin *in* Martin & Dean, p. 36, Pl. 12:1, 2. □1989 *Cristallinium cambriense* (Slaviková 1968) Vanguestaine 1978 – Mette, p. 2, Pl. 1:4, 7. □1990 *Cristallinium cambriense* (Slaviková) Vanguestaine 1978 – Tongiorgi & Ribecai, Pl. 1:3. □1990 *Cristallinium ovillense* (Cramer & Diez, 1972a) comb.nov. – Fensome *et al.*, p. 161. □1991 *Cristallinium cambriense* (Slaviková) Vanguestaine 1978 – Di Milia, Pl. 1:12–13. □1991 *Cristallinium cambriense* (Slaviková) Vanguestaine, 1978 – Albani *et al.*, pp. 266–267, Pl. 1:3–5. □1993 *Cristallinium ovillense* (Cramer & Diez) Fensome *et al.* 1990 – Ribecai & Vanguestaine, p. 55, Pl. 1:5. □1993 *Cristallinium ovillense* (Cramer & Diez) Fensome, Williams, Sedley Barss, Freeman and Hill 1990 – Fombella *et al.*, p. 230, Pl. 2:2. □1994 *Cristallinium cambriense* (Slaviková) Vanguestaine, 1978 – Martin in Young *et al.*, pp. 353–355, Fig. 11i. □1994 *Cymatiosphaera ovillensis* Cramer & Diez, 1972 – Martin in Young *et al.*, pp. 353–355, Fig. 10j, q. □1995 *Cristallinium cambriense* (Slaviková 1968) Vanguestaine 1978 – Moczydłowska & Crimes, p. 119–120, Pl. 2E, F.

   See also a comprehensive synonymy by Welsch (1986, p. 75) and Albani *et al.* (1991, pp. 266, 268).

Material. – Twenty-eight well-preserved specimens.

Description. – Vesicles oval to subpolygonal, originally globular, divided by low costae into pentagonal and hexagonal fields. The surface of the polygonal fields and edges of the costae are chagrinate to granular.

Dimensions. – $N = 15$. Vesicle diameter 23–45 µm, diameter of polygonal fields 7–16 µm.

Remarks. – *Cristallinium cambriense* was reported from the Umbria Pipeta Formation in Sierra Morena, SW Spain, and interpreted as Middle Cambrian in age (Mette 1989). Because of the lack of other fossils, the relative age of the formation was only inferred from the acritarch assemblage (including additionally *C. randomense*?, *Eliasum llaniscum*, *E. asturicum* and *Timofeevia phosphoritica*) to be time-equivalent to the *Paradoxides paradoxissimus* Superzone. This assumption was based on the incomplete stratigraphic range of *E. llaniscum*, believed to be limited to the Middle Cambrian *A. oelandicus* and *P. paradoxissimus* Superzones (Mette 1989, p. 2); in fact, it seems to extend to Upper Cambrian (Moczydłowska & Crimes 1995, and present data). Aside from this, the presence of *C. randomense* in the assemblage, if correctly identified, would suggest a Late Cambrian age (Martin & Dean 1981, 1988; Welsch 1986) for the Umbria Pipeta Formation.

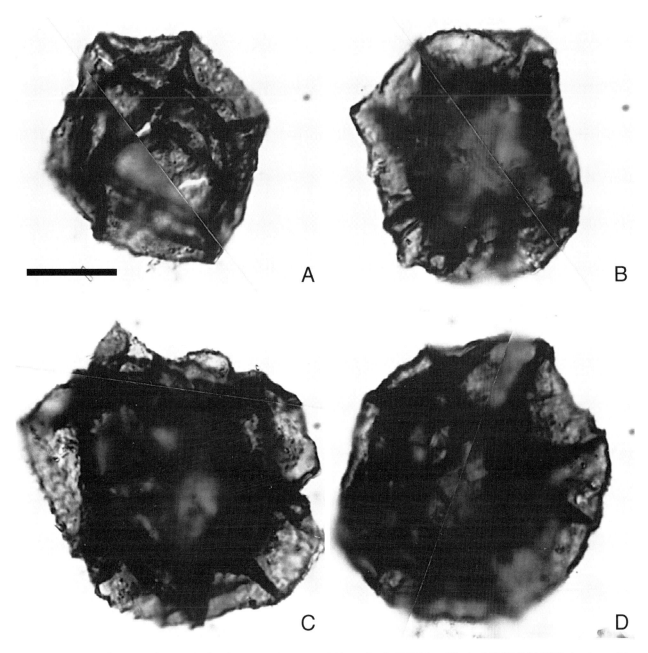

Fig. 23. □A–D. *Cristallinium cambriense* (Slaviková) Vanguestaine. A, PMU-Pl.24-R/40; B, PMU-Pl.25-N/35/1; C, PMU-Pl.26-H/27, showing vesicles subpolygonal in outline and divided by low costae into polygonal fields. The granular sculpture occurs both on the surface of the fields and on the edges of the costae. D, PMU-Pl.27-Z/28/4; specimen with more circular outline. All specimens from the Sosnowiec IG-1 borehole, depth 3174.0–3174.7 m, Sosnowiec formation, lower Upper Cambrian. Scale bar in A equals 12 μm for all micrographs.

Present record. – The Upper Silesia area, Goczałkowice IG-1 borehole, Goczałkowice formation at a depth of 2766.8–2771.1 m, Middle Cambrian, *Acadoparadoxides oelandicus* Superzone; Sosnowiec IG-1 borehole, Sosnowiec formation at an interval of 3174.0–3372.3 m, Middle Cambrian, *Paradoxides paradoxissimus* Superzone to lower Upper Cambrian.

Occurrence and stratigraphic range. – The species has a widespread geographic distribution, ranging from Middle Cambrian to Tremadocian (Fig. 14). A detailed compilation of occurrences is provided by Welsch (1986, p. 76) and Albani *et al.* (1991, p. 268). Additionally: Russia, Moscow Syneclise, Tolbukhino 1 borehole, Sablin Formation, upper Middle – lower Upper Cambrian (Volkova 1980) and St. Petersburg area, locality at River Tosna and Zariechje borehole, Izhora Formation, Upper Cambrian (Jankauskas 1980). Lithuania, Aukshten–Paneryaj borehole, Lakaj Formation, Middle Cambrian (Jankauskas 1980). Italy, Sardinia, Arburese Unit, Upper Cambrian, Lower Tremadocian (Pittau 1985); central Sardinia, localities at River S. Giorgio and River Araxisi, Solanas Sandstone Formation, Upper Cambrian (Di Milia 1991). Czech Republic, Bohemian Massif at Jince, Jince Formation, Middle Cambrian, *Eccaparadoxides pusillus* zone (Vavrdová 1976), and Vinice Hill near Jince, Jince Formation, Middle Cambrian, *Onymagnostus hybridus* Zone (=*Ptychagnostus atavus* and *Tomagnostus fissus* Zone) (Fatka 1989). SW Spain, Sierra Morena, Province Huelva, Cañaveral de Leon, Umbria Pipeta Formation, Middle Cambrian (Mette 1989), but probably it is Upper Cambrian (see comments above). Libya, Rhadames Basin, A1-70 borehole and Tunisia, Tt1 borehole, Sidi–Toui Formation, upper Middle Cambrian to lower Upper Cambrian (Albani *et al.* 1991). Belgium, Stavelot Massif and Rocroi Massif, Revin Group, Middle and Upper Cambrian (Vanguestaine 1978, 1992; Ribecai & Vanguestaine 1993). NW Wales, St. Tudwal's Peninsula and St. Tudwal's Island East, Ceiriad Formation, Nant-y-big Formation, Middle Cambrian, *Acadoparadoxides oelandicus* and *Paradoxides paradoxissimus* Superzones; Maentwrog Formation and Ffestiniog Flags Formation, Upper Cambrian (Young *et al.* 1994). Ireland, Co. Wexford, the Booley Bay Formation, upper part of Upper Cambrian (Moczydłowska & Crimes 1995).

Occurrence of the species in the presumably upper Lower Cambrian in Sweden and Finland (Moczydłowska & Vidal 1986; Hagenfeldt 1989b; Eklund 1990) is uncertain because of the difficulties in recognizing the Lower–Middle Cambrian boundary in the strata concerned. The record of the species from the upper File Haidar Formation in Sweden and upper Söderfjärden Formation in Finland (Hagenfeldt 1989b) is not proven to be of early Cambrian age (see under 'Re-evaluation of stratigraphic ranges').

The occurrence of a few species of *Cymatiosphaera* that are synonymous with *C. cambriense* (see under 'Taxonomy') in alleged Lower Cambrian strata (Jankauskas 1976; Volkova *et al.* 1979) was revised later by Jankauskas (1980) and regarded as Middle–Upper Cambrian. Consequently, the lower range of *C. cambriense* as Lower Cambrian cited by Welsch 1986 (p. 76) and Fensome *et al.* 1990 should be abandoned.

Cristallinium ovillense (Cramer & Diez, 1972) Martin, 1981

Fig. 24B

Synonymy. – □1972 *Zonosphaeridium* sp. 1 – Cramer, pp. 1–4, Pl. 1:4, 6. □1972 *Zonosphaeridium ovillensis* Cramer & Diez (New Species) – Cramer & Diez, p. 44, 47, Pl. 2:5, 8, 11. □1976 *Zonosphaeridium ovillensis* Cramer & Diez 1972 – Eisenack, Cramer & Diez, p. 849. □1978 *Zonosphaeridium* sp., en: Cramer & Diez – Fombella, Pl. 3:16, 17. □1979 *Zonosphaeridium ovillensis* Cramer & Diez 1972 – Fombella, Pls. 2:25; 4:60, 61. □1981 *Zonosphaeridium* cf. *ovillensis* Cramer & Diez – Erkmen & Bozdoğan, p. 56, Pl. 2:12. □1981 *Cristallinium ovillense* (Cramer & Diez de Cramer) comb.nov. – Martin *in* Martin & Dean, pp. 17–18, Pl. 3:16. □1985 *Zonosphaeridium ovillensis* Cramer & Diez 1972 – Pittau, p. 201, Pl. 7:1, 2, 4. □1986 *Cristallinium ovillense* (Cramer & Diez 1972) – Welsch, pp. 76–77, Pl. 2:25–28. □1986 *Cristallinium ovillense* (Cramer & Diez de Cramer) comb.nov. – Fombella, Pl. 3:15. □1986 *Cristallinium ovillense* Martin *in* Martin & Dean 1981 – Fombella, Pl. 3:18. □1990 *Zonosphaeridium* cf. *Z. ovillensis* Cramer et Diez, 1972 – Volkova, p. 90, Pl. 2:2, 5. □1990 *Cristallinium deceptum* sp.nov. – Fensome *et al.* p. 160. □1991 *Cristallinium ovillense* (Cramer & Diez) Martin, 1981 – Di Milia, p. 134, 136, Pl. 1:10–11. □1991 *Cristallinium* sp. aff. *C. ovillense* (Cramer & Diez) Martin 1981 – Albani *et al.*, pp. 268–269, Pl. 1:6. □1993 *Cristallinium deceptum* Fensome, Williams, Sedley Brass, Freeman y Hill, 1990 – Fombella *et al.*, p. 230, Pl. 3:2.

Material. – Eight well-preserved specimens.

Description. – Vesicles circular in outline, originally spherical, divided by low costae into polygonal fields forming a dense net. The polygonal fields are pentagonal and hexagonal, small and regular. The surfaces of wall and costae are smooth. The excystment by rupture of the wall is much smaller than the diameter of the vesicle.

Dimensions. – $N = 2$. Vesicle diameter 29–36 μm, diameter of the polygonal fields 4–6 μm.

Remarks. – The genus *Zonosphaeridium* Timofeev, 1956, *ex* Timofeev, 1959, was not validly published, because the type species was not indicated. The species *Z. ovillensis* Cramer & Diez, 1972 was transferred to *Cristallinium* by Martin (*in* Martin & Dean 1981). The name *Cristallinium deceptum* sp.nov., newly proposed by Fensome *et al.* (1990, pp. 160–161) to replace *C. ovillense* (Cramer & Diez, 1972) Martin, 1981, is rejected (see comments above).

Cristallinium sp. aff. *C. ovillense* (Cramer & Diez, 1972) Martin, 1981, reported by Albani *et al.* (1991), is tentatively included in the synonymy of *C. ovillense*. It has granular ornamentation on the vesicle surface and on the

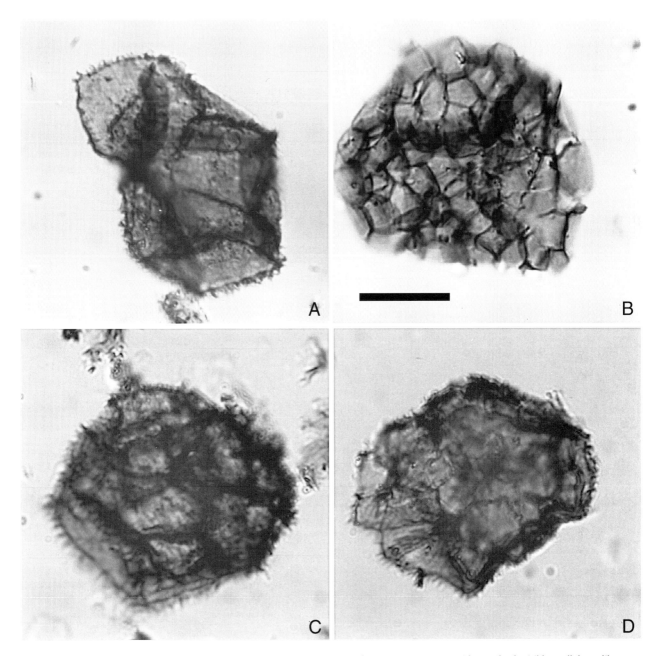

Fig. 24. □A, C–D. *Cristallinium randomense* (Martin) Martin. A, PMU-Pl.28-F/42, fragmentary specimen, with very clearly visible small thorn-like processes on the edge of septa, dividing the vesicle on polygonal fields, and granulate surface of the fields; C, PMU-Pl.29-K/46; D, PMU-Pl.30-N/38, low focus of specimens displaying small processes. □B. *Cristallinium ovillense* (Cramer & Diez) Martin. PMU-Pl.31-D/36/3; showing dense net of low costae and smooth surface of both vesicle wall and costae. The simple rupture of the vesicle in lower half of specimen probably represents the excystment structure. All specimens from the Sosnowiec IG-1 borehole, depth 3174.0–3174.7 m, Sosnowiec formation, lower Upper Cambrian. Scale bar in B equals 12 μm for all micrographs.

crests, contrary to the smooth vesicle wall of *C. ovillense* (Cramer & Diez 1972; Martin & Dean 1981; Pittau 1985). Welsch (1986) suggested that specimens with the ornamentation might be separated from this species. However, the specimen of *C. ovillense* illustrated by Martin (*in* Martin & Dean 1981, Pl.3:16) does not display ornamentation, contrary to the suggestion by Welsch (1986, p. 77)

and Albani *et al.* (1991, p. 269), and it need not be transferred to another species.

Present record. – The Upper Silesia area, Sosnowiec IG-1 borehole, Sosnowiec formation, in the interval of 3174.0–3205.5 m, Middle Cambrian, *Paradoxides forchhammeri* Superzone to lower Upper Cambrian.

Occurrence and stratigraphic range. – Spain, the Cantabrian Mountains at Láncara de Luna, the Oville Formation, upper Middle Cambrian (Cramer & Diez 1972) and other localities with the Oville Formation referred to as upper Middle Cambrian and Upper Cambrian – Tremadoc by Fombella (1978, 1979). The latter records from the Oville Formation referring to the Upper Cambrian – Tremadoc are taken here with the reservation, as such stratigraphic extension of the formation has never been documented (see under 'Range'). SE Turkey, Mardin–Derik area, Sosink Formation, Middle Cambrian (Erkmen & Bozdoğan 1981), here considered equivalent to the *Paradoxides paradoxissimus* Superzone. Canada, Newfoundland, Random Island, Elliott Cove Formation, Upper Cambrian, interval between the *Olenus* Zone and *Parabolina spinulosa* Zone (Martin & Dean 1981). Italy, Sardinia, Arburese Unit, Tremadocian, species regarded as redeposited (Pittau 1985), and central Sardinia, River S. Giorgio and River Araxisi, Solanas Sandstone Formation, Upper Cambrian (Di Milia 1991). Norway, Finnmark, Digermul Peninsula, Kistedal Formation, Middle Cambrian, *Paradoxides paradoxissimus* Superzone (Welsch 1986). Russia, Kaliningrad area, Veselovsk-5 borehole, Veselovsk Formation, Middle Cambrian (Volkova 1990). Libya, Rhadames Basin, A1-70 borehole, Sidi-Toui Formation, lower Upper Cambrian (Albani *et al.* 1991).

Cristallinium randomense (Martin, 1981) Martin, 1988

Fig. 24A, C–D

Synonymy. – □1981 *Cristallinium randomense* sp.nov. – Martin *in* Martin & Dean, p. 18, Pls. 3:2, 10, 12, 17, 20, 24, 26; 6:4, 6. □1982 *Cristallinium randomense* – Martin, Pl. 1:18. □1983 *Cristallinium* sp. – Volkova, Pl. 2:3, 5. □1983 *Cristallinium* aff. *ovillensis*; Volkova, Pl. 2:6. □1986 *Cristallinium randomense*; Welsch, p. 77, Pl. 5:12–17. □1988 *Cristallinium randomense* Martin *in* Martin & Dean 1981 emend. Martin herein – Martin & Dean, pp. 36–37, Pl. 13:1–17. □1990 *Cristallinium baculatum* Volkova, sp.nov. – Volkova, p. 60, Pl. 14:12, 13. □1990 *Cristallinium randomense* Martin 1981 emend. Martin 1988 – Tongiorgi & Ribecai, Pl. 1:2. □1992 *Cristallinium* sp. – Paalits, Pl. 4:5. □*non* 1993 *Cristallinium randomense* (Martin) Fensome *et al.* 1990 – Ribecai & Vanguestaine, p. 55, Pl. 1:6. □1995 *Cristallinium randomense* – Moczydłowska & Crimes, p. 120, Pl. 2A–D.

Material. – Twenty-three fairly well-preserved specimens.

Description. – Vesicles polygonal in outline, originally globular, divided by low septa into pentagonal and hexagonal fields. The surface of the polygonal fields is chagri-nate to granulate. The edges of the septa bear numerous short and solid processes that may also occur occasionally within the polygonal fields. The shape of the processes varies from conical, thorn-like to cylindrical. Their tips are simple sharp, truncated or branching.

Dimensions. – $N=7$. Vesicle diameter 27–46 µm, diameter of polygonal fields 7–15 µm, length of processes less than 2 µm.

Remarks. – The specimen attributed to *Cristallinium randomense* (Martin) Fensome *et al.* 1990 by Ribecai & Vanguestaine (1993, Pl. 1:6) represents the species *Cristallinium pilosum* Golub & Volkova, 1985 (*in* Volkova & Golub 1985).

The specimens in the present material possess processes that are shorter than the size range ascribed to the species by Martin (*in* Martin & Dean 1981, 1988). However, because of the various states of preservation, most of the specimens of *C. randomense* illustrated with the original and emended diagnoses (Martin & Dean 1981, 1988) have shorter processes as well, or may have only locally preserved ornamentation on a particular specimen. One of the diagnostic features of *C. randomense* that is observable on specimens in the original material irrespective of the state of preservation (Martin & Dean 1981, 1988; and personal examination of the collection) is the presence of solid, stubby process bases, 1–2 µm long (e.g., 'squat projections', Martin & Dean 1988). The filamentous spines (1–3) extending from these bases and having maximum length 5 µm are less frequently preserved. In most instances, the preserved portion of the processes is around 2 µm, as in the present material.

Present record. – The Upper Silesia area, Sosnowiec IG-1 borehole, Sosnowiec formation at a depth of 3174.0–3174.7 m; lower Upper Cambrian.

Occurrence and stratigraphic range. – Canada, Newfoundland, Random Island and Manuels River, Elliot Cove Formation, Upper Cambrian (*Parabolina spinulosa* Zone to *Acerocare* Zone), and Clarenville Formation, Tremadoc (Martin & Dean 1981, 1988; Martin 1982). Belgium, Stavelot Massif, Rocroi Massif and France, the Ardennes – Upper Cambrian (Meilliez & Vanguestaine 1983). Norway, Finnmark, Digermul Peninsula, Kistedal Formation, Upper Cambrian (*Agnostus pisiformis* Zone to *Acerocare* Zone) and Berlogaissa Formation, Tremadoc (*Obolus* Zone and *Clonograptus tenellus* Zone) (Welsch 1986). Estonia, Tallinn area, Upper Cambrian (Volkova 1990), and M-72 borehole, Tsitre Formation (='Obolus sandstone'), Upper Cambrian (Paalits 1992). Ireland, Co. Wexford, Booley Bay Formation, upper part of Upper Cambrian (Moczydłowska & Crimes 1995).

Genus *Cymatiosphaera* O. Wetzel, 1933 *ex* Deflandre, 1954

Type species. – *Cymatiosphaera radiata* O. Wetzel, 1933b, p. 27, Pl. 4:8, *emend.* Sarjeant 1985, pp. 161–162; Baltic area, Late Cretaceous.

Cymatiosphaera cramerii Slaviková, 1968

Fig. 25D–H

Synonymy. – □1968 *Cymatiosphaera crameri* n.sp. – Slaviková, pp. 200–201, Pl. 1:8. □1983 *Cymatiosphaera crameri* Slaviková, 1968 – Martin & Dean, Pl. 43.2, Fig. 1. □1984 *Cymatiosphaera crameri* Slaviková, 1968 – Martin & Dean, Pl. 57.1, Fig. 8. □1989 *Cymatiosphaera cramerii* Slaviková, 1968 – Fatka, p. 365, Pl. 1:9.

Material. – Twelve specimens in different states of preservation.

Description. – Vesicles circular, oval or slightly polygonal in outline, spherical before compaction, with surface subdivided into polygonal fields by numerous high ridges. The ridges arise perpendicularily from the vesicle surface. The number of ridges, and therefore the number of polygonal fields, is variable. The fields are pentagonal and hexagonal.

Dimensions. – $N = 12$. Overall diameter of vesicles 22–45 μm, diameter of fields 5–7 μm.

Remarks. – The original spelling of the specific epithet 'crameri' is modified to 'cramerii' following the recomendations of The International Code of Botanical Nomenclature (1988) to form a new specific epithet commemorating a person (Stearn 1983, p. 295).

Morphotypes of *Cymatiosphaera* similar to the present material have been recorded in the middle Cambrian assemblage from the Bohemian Massif in the Czech Republic (Slaviková 1968). They were assigned to various species, *Cymatiosphaera crameri* Slaviková, 1968, *C.* cf. *nebulosa* (Deunff) and *C.* cf. *mirabilis* Deunff, though they do not reveal recognizable morphological dissimilarities. Most likely, all are conspecific with *C. cramerii*.

Present record. – The Upper Silesia area, Sosnowiec IG-1 borehole, Sosnowiec formation at depths of 3365.2–3372.3 m and 3210.0–3212.0 m, Middle Cambrian, *Acadoparadoxides oelandicus* and *Paradoxides paradoxissimus* Superzones, respectively; 3174.0–3174.7 m, lower Upper Cambrian.

Occurrence and stratigraphic range. – Czech Republic, Bohemian Massif, Brdy Mountains at Stašice, Těně Sš-III borehole, Jince Formation, Middle Cambrian, *Ellipsocephalus hoffi* Zone (Slaviková 1968); Bohemian Massif,

Jince and Stryje area, Middle Cambrian, *Eccaparadoxides pusillus*, *Paradoxides gracilis* and *Hydrocephalus lyelli* Zones (Vavrdová 1976); Bohemian Massif, Přibram–Jince Basin, Jince Formation, Middle Cambrian, *Onymagnostus hybridus* Zone (=*Tomagnostus fissus* and *Ptychagnostus atavus* Zone) (Fatka 1989). Canada, Newfoundland, Manuels River and Random Island, Chamberlains Brook Formation, Middle Cambrian, *Paradoxides bennettii* Zone (Martin & Dean 1983, 1984).

Cymatiosphaera postae (Jankauskas, 1976) Jankauskas, 1979

Fig. 26C, F

Synonymy. – □*non* 1933 *Cymatiosphaera radiata* O. Wetzel 1933b (p. 27, Pl. 4:8) emend. Sarjeant, 1985 (pp. 161–162). □illegitimate 1976 *Cymatiosphaera radiata* Jankauskas, sp. nov – Jankauskas & Posti, pp. 148–149, Figs 2, 3. □1979 *Cymatiosphaera postii* Jankauskas, nom.nov. – Jankauskas in Volkova *et al.*, p. 26, Pl. 14:1–3. □1982 *Cymatiosphaera postii* Jankauskas 1979 – Downie, p. 264, Fig. 9h. □1991 *Cymatiosphaera postii* (Jankauskas, 1976) Jankauskas, 1979 – Moczydłowska, p. 52, Pl. 9H–I.

Material. – Six well-preserved specimens.

Description. – Vesicles circular to oval in outline, originally spherical, having high ridges which divide the surface of the vesicle into pentagonal and hexagonal fields. Six to eight ridges are usually observed on the outline of the vesicle

Dimensions. – $N = 14$. Overall vesicle diameter 23–45 μm, diameter of central body 15–20 μm, diameter of polygonal fields 10–12 μm, height of ridges 6–10 μm.

Remarks. – *Cymatiosphaera radiata* Jankauskas, 1976 (in Jankauskas & Posti 1976) was a junior homonym of *C. radiata* O. Wetzel, 1933, and thus an illegitimate name for the new species. It was subsequently renamed as *C. postii* (Jankauskas in Volkova *et al.* 1979). The specific epithet *postii* is corrected to *postae* because it derives from the feminine gender name of Erika Posti (Jankauskas & Posti 1976) and ought to carry the suffix -*ae* (ICBN 1988, Stearn 1983).

Present record. – The Upper Silesia area, Sosnowiec IG-1 borehole, Sosnowiec formation at a depth of 3365.2–3372.3 m, Middle Cambrian, *Acadoparadoxides oelandicus* Superzone.

Occurrence and stratigraphic range. – Lithuania, Lower Cambrian, Vergale and Rausve horizons, and Middle Cambrian Kibartai horizon; Estonia and Latvia, Lower Cambrian, Rausve horizon (Volkova *et al.* 1979). Sweden, Västergötland, File Haidar Formation, *Mickwitzia*

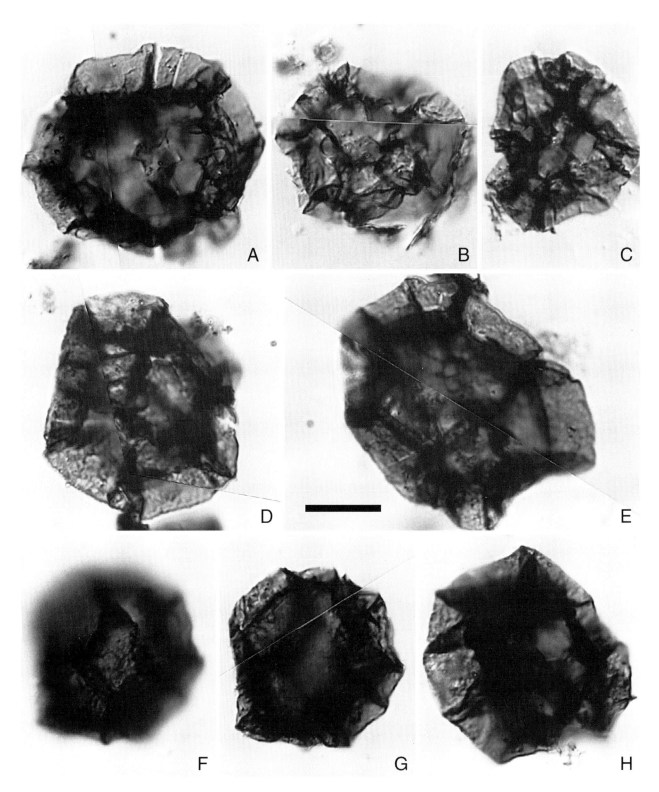

Fig. 25. □A–C. *Cymatiosphaera* sp.; A, PMU-Pl.38-R/47/3; B, PMU-Pl.39-W/48; C, PMU-Pl.40-E/35; specimens with large vesicles, low ridges and irregular polygonal fields. □D–H. *Cymatiosphaera cramerii* Slaviková. D, PMU-Pl.41-K/31/2; E, PMU-Pl.42-X/46; F–G, PMU-Pl.43-V/39/3, at different focus levels; H, PMU-Pl.44-D/31; showing varying numbers of polygonal fields. All specimens from the Sosnowiec IG-1 borehole, Sosnowiec formation; A, depth 3174.0–3174.7 m, lower Upper Cambrian; B–H, depth 3365.2–3372.3 m, Middle Cambrian, *Acadoparadoxides oelandicus* Superzone. Scale bar in E equals 12 μm for all micrographs.

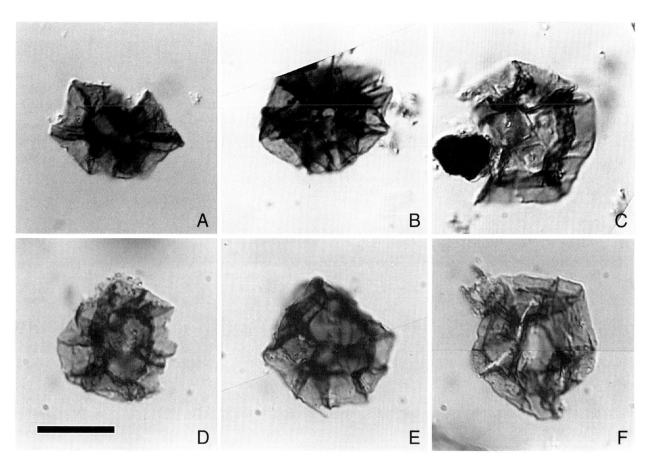

Fig. 26. □A–B. *Cymatiosphaera pusilla* n.sp. A, PMU-Pl.32-U/39/3, holotype; B, PMU-Pl.33-S/29, showing hexagonal vesicle outline and small central body with high membranaceous ridges. □C, F, *Cymatiosphaera postae* (Jankauskas) Jankauskas. C, PMU-Pl.34-Y/37; F, PMU-Pl.35-Q/42; specimens with regular outline of vesicle and penta- or hexagonal fields formed by ridges on the vesicle surface. □D–E. *Cymatiosphaera* sp. A. D, PMU-Pl.36-M/32/2; E, PMU-Pl.37-L/37, with radial ridges pattern on the vesicle outline. All specimens from the Sosnowiec IG-1 borehole, Sosnowiec formation; A, C, F, depth 3365.2–3372.3 m, Middle Cambrian, *Acadoparadoxides oelandicus* Superzone; B, D, E, depth 3174.0–3174.7 m, lower Upper Cambrian. Scale bar in D equals 12 μm for all micrographs.

Sandstone Member, Lower Cambrian, *Holmia kjerulfi* Zone (Moczydłowska & Vidal 1986, Moczydłowska 1991); Island of Gotland, Grötlingbo-1 borehole, and Gotska Sandön Island, Gotska Sandön borehole, File Haidar Formation and 'oelandicus beds'; the Gulf of Bothnia, Finngrundet borehole, Söderfjärden Formation, Lower Cambrian, *Holmia kjerulfi* and *Protolenus* Zones, and Middle Cambrian, *Acadoparadoxicus* Superzone; south-central Sweden, Närke at Kvarntorp, 'Oelandicus beds', Middle Cambrian, *Acadoparadoxides oelandicus* Superzone (Hagenfeldt 1989 a). Finland, Vaasa area, Söderfjärden-3 borebole, Söderfjärden Formation, Lower Cambrian, *Holmia kjerulfi* and *Protolenus* Zones, and Middle Cambrian, *Acadoparadoxides oelandicus* Superzone (Hagenfeldt 1989 a). Norway, Lake Mjøsa region, Lower Cambrian, *Holmia kjerulfi* Zone (Moczydłowska & Vidal 1986; Moczydłowska 1991). Scotland, Fucoid Beds, Lower Cambrian (Downie 1982). East Greenland, Ella Ø, Bastion Formation, Lower Cambrian (Downie 1982), North Greenland, Peary Land, Buen Formation, Lower Cambrian, time equivalent to *Holmia kjerulfi* and *Protolenus* Zones (Vidal & Peel 1993). Poland, Lublin Slope of the East European Platform, Kaplonosy and Radzyń formations, Lower Cambrian, *Holmia kjerulfi* and *Protolenus* Zones (Moczydłowska 1991).

In Russia the species was reported from the Izhora beds, previously referred to the Lower? Cambrian (Volkova *et al.* 1979) and subsequently re-assessed as Upper Cambrian (Jankauskas 1980; Volkova 1990; Moczydłowska & Crimes 1995). However, the occurrence of the species in the Izhora beds has never been documented by photomicrographs or confirmed after its mention in 1979, and thus it is here treated as uncertain.

Cymatiosphaera pusilla n.sp.

Fig. 26A, B

Holotype. – Specimen PMU-Pl.32-U/39/3; Fig. 26A.

Synonymy. – □1979 *Cymatiosphaera* div. sp. – Volkova *et al.*, Pl. 15:4.

Derivation of name. – From Latin *pusillus* – small, referring to the overall dimensions of the vesicle.

Locus typicus. – The Upper Silesia area, Sosnowiec IG-1 borehole (Figs. 2 and 3).

Stratum typicum. – Alternating mudstones and sandstones in the Sosnowiec formation at a depth of 3365.2–3372.3 m; Middle Cambrian, *Acadoparadoxides oelandicus* Superzone.

Material. – Six well-preserved specimens.

Diagnosis. – Vesicles hexagonal to polygonal in outline, having a small central body divided into polygonal fields by high membranous ridges, which have concave margins and arise perpendicular to the surface of the vesicle. Number of ridges observed along the outline 6–8.

Dimensions. – $N=4$. Overall diameter of vesicles 14–27 μm, diameter of central body 8–10 μm, height of ridges 7–10 μm.

Remarks. – *C. pusilla* n.sp. differs from other species of *Cymatiosphaera* by its small overall dimensions and relatively small central body, and by having an hexagonal outline of the vesicle and concave margins of the ridges.

Present record. – The Upper Silesia area, Sosnowiec IG-1 borehole, Sosnowiec formation in the interval of 3174.0–3372.3 m, Middle Cambrian, *Acadoparadoxides oelandicus* Superzone to lower Upper Cambrian.

Occurrence and stratigraphic range. – The Ukraine, Volhyn region, Shatsk borehole, Svityaz Formation, uppermost Lower Cambrian, Rausve horizon (Volkova *et al.* 1979, p. 197)

Cymatiosphaera sp. A

Fig. 26E

Material. – Three well-preserved specimens.

Description. – Small subpolygonal vesicles consisting of a spherical central body divided into polygonal fields by relatively low ridges. The wall of the vesicle and the ridges are thin and smooth. The ridges observed on the outline of the vesicle are very regular and arise perpendicular to the surface of the central body.

Dimensions. – $N=3$. Overall diameter 23–25 μm, diameter of central body around 15 μm, diameter of fields 7–8 μm, height of ridges ca. 5 μm.

Remarks. – The species differs from *Cymatiosphaera postae* (Jankauskas, 1976) Jankauskas, 1979, by having more circular fields, a more spherical outline of the vesicle, and thinner ridges.

Present record. – The Upper Silesia area, Sosnowiec IG-1 borehole, Sosnowiec formation at a depth of 3174.0–3174.7 m, lower Upper Cambrian.

Cymatiosphaera sp.

Fig. 25A–C.

Material. – Twenty-eight species in various states of preservation.

Present record. – The Upper Silesia area, the Goczałkowice IG-1 borehole, Goczałkowice formation at a depth of 2766.8–2771.1 m, Middle Cambrian, *Acadoparadoxides oelandicus* Superzone; the Sosnowiec IG-1 borehole, Sosnowiec formation in the interval of 3174.0–3423.8 m, Middle Cambrian, *Acadoparadoxides oelandicus* Superzone to lower Upper Cambrian.

Genus *Duplisphaera* n.gen.

Type species. – *Duplisphaera luminosa* (Fombella, 1978) n.comb. (=*Cymatiosphaera luminosa*), Fombella, 1978, pp. 251–252, Pl. 2:9; Spain, Cantabrian Mountains, Province of León, Oville Formation, upper Middle Cambrian.

Derivation of name. – From Latin *duplus* – double, and *sphaera* – sphere, referring to the double-layered spherical wall of the vesicle.

Diagnosis. – Acritarchs with doubled-walled vesicles consisting of an inner body bearing radially arranged processes that support the outer spherical membrane enclosing the inner body and processes. The inner body and processes are more dense in appearance than the outer membrane, which is thin and translucent. The processes are solid, in the shape of spokes with variable shaped tips supporting the outer membrane.

Remarks. – The new genus differs from *Fimbriaglomerella* Loeblich & Drugg, 1968, by lacking filmy muri stretching between the processes and forming polygonal luminae on the outher wall. It is smaller than *Cymatiosphaeroides* Knoll, 1983 (*in* Knoll & Calder 1983), and have more sparsely distributed processes.

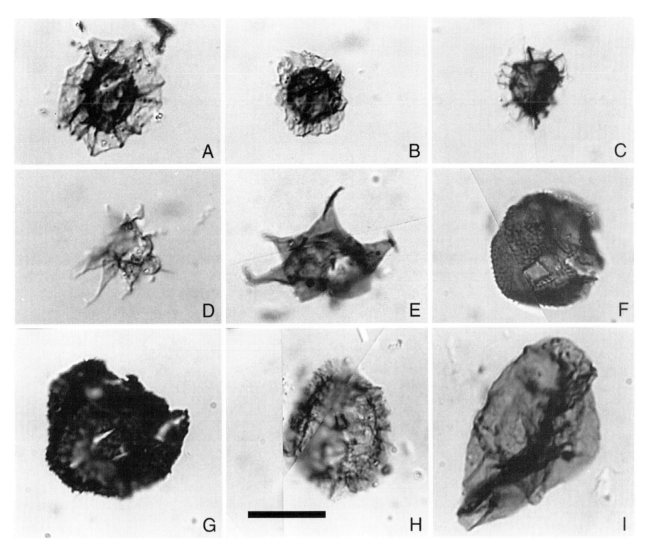

Fig. 27. □A–C. *Duplisphaera luminosa* (Fombella) n.comb. A, PMU-Pl.45-H/35/1, showing double-walled vesicle with radially arranged processes arising from the inner body and supporting the outher membrane; B, PMU-Pl.46-L/45/4, processes are less clearly visible because of the thick outer membrane; C, PMU-Pl.47-O/29/2, showing processes with differing tips, blunt and distally widened. □D–E. *Estiastra minima* Volkova. D, PMU-Pl.48-S/31, processes with elongated thin distal portions; F, PMU-Pl.49-Z/29; showing the stellate shape of the vesicle with less clear differentiation of processes from the central body. □F. *Lophosphaeridium latviense* (Volkova) n.comb., PMU-Pl.50-O/38/3, with diagnostic very regular and evenly distributed ornamentation elements shaped as short solid thorns. □G. *Lophosphaeridium variabile* Volkova. PMU-Pl.51-R/28, showing thick-walled vesicle with coarse granular sculpture on the wall and rim of excystment structure, a simple circular opening. □H. *Pterospermella vitalis* Jankauskas. PMU-Pl.52-X/43/2, with short radial rod-like elements supporting a transparent membrane in the equatorial plane of the vesicle. □I. *Leiosphaeridia* sp. PMU-Pl.53-E/39/3. All specimens from the Sosnowiec IG-1 borehole, Sosnowiec formation. A, B, depth 3174.0–3174.7 m, lower Upper Cambrian; C, D, H, I, depth 3365.2–3372.3 m; E, F, G, depth 3403.5–3407.2 m, Middle Cambrian, *Acadoparadoxides oelandicus* Superzone. Scale bar in H equals 12 μm for all micrographs.

Duplisphaera luminosa (Fombella, 1978) n.comb.

Fig. 27A–C

Synonymy. – □1978 *Cymatiosphaera luminosa* Fombella 1978, n.sp. – Fombella, pp. 251–252, Pl. 2:9. □*non* 1982 *Cymatiosphaera* cf. *luminosa* Fombella 1978 – Downie, p. 265, Fig. 9i–j. □1987 *Cymatiosphaera luminosa* Fombella, 1978 – Fombella, Pl. 1:13.

Material. – Six well-preserved specimens.

Description. – Double-walled spherical vesicles having an inner body bearing radially arranged processes that may have blunt or slightly swollen and rounded tips supporting the outer membrane. The outer membrane (wall) is thin and translucent, whereas the wall of the inner body and processes is more dense. The processes are solid and rod-like.

Dimensions. – $N = 6$. Overall vesicle diameter 13–18 μm, inner body diameter 6–13 μm, height of processes 3–5 μm, number of processes observed on the outline of the inner body 11–13.

Remarks. – The holotype of *Cymatiosphaera luminosa* Fombella, 1978 (Fombella 1978, pp. 251–252, Pl. 2:9), as described and illustrated, does not belong to the genus *Cymatiosphaera*. It lacks ridges stretching from the surface of the vesicle, and the polygonal fields formed by them that are diagnostic for *Cymatiosphaera* (Eisenack *et al.* 1973). A further significant dissimilarity is the double-walled vesicle.

The stratigraphic range of *C. luminosa* was not clearly assessed in the original descritpion, nor was the stratum typicum of the holotype indicated. The section at the type locality in Corniero (Fombella 1978) was never measured, and the only information about the stratum typicum is that it is the Oville Formation. By a general estimate, the Oville Formation in this area is 106–190 m thick and spans an interval of the Middle Cambrian (Aramburu *et al.* 1992). The species was shown to occur within a very narrow range at the base of zone 2, referred to the upper part of the Middle Cambrian (Fombella 1978, Fig. 2). In a later publication, however, Fombella (1987) attributed *C. luminosa*, recorded in the Oville Formation in the same region at Vozmediano, to the Upper Cambrian. The lack of faunal evidence and information about the relative occurrence of the acritarch species within the measured geological succession prevents establishment of a more precise stratigraphic range of the species in the Oville Formation than Middle Cambrian (see under 'Stratigraphic ranges').

Present record. – The Upper Silesia area, Sosnowiec IG-1 borehole, Sosnowiec formation at depths of 3365–3372.3 m and 3210.0–3212.0 m, 3174.0–3174.7 m; Middle Cambrian, *Acadoparadoxides oelandicus* and *Paradoxides paradoxissimus* Superzones, and lower Upper Cambrian, respectively.

Occurrence and stratigraphic range. – Spain, the Cantabrian Mountains at Corniero, the Oville Formation, upper part of Middle Cambrian (Fombella 1978), and at Vozmediano, the Oville Formation, Upper Cambrian (Fombella 1987). The latter occurrence is regarded here as Middle Cambrian (see under 'Stratigraphic ranges').

Fensome *et al.* (1990, p. 172) misquoted the age of *Cymatiosphaera luminosa* Fombella, 1978 as 'Early Cambrian'; it should be late Middle Cambrian (Fombella 1978, pp. 248, 252).

Genus *Eliasum* Fombella, 1977

Type species. – *Eliasum llaniscum* Fombella, 1977, p. 118, Pl. 1:6; Spain, the Cantabrian Mountains, the Oville Formation, lower part of Middle Cambrian.

Eliasum llaniscum Fombella, 1977

Fig. 28A–D

Synonymy. – □1966 *Leiosphaeridia* sp – Vavrdová, Pl. 2:2. □1972 *Leiosphaeridia* sp. 2 – Cramer & Diez, Pl. 2:9. □1976 *Leiosphaeridia* sp. – Vavrdová, pp. 60–61, Pl. 1:4. □1977 *Eliasum llaniscum* Fombella n.sp. – Fombella, p. 118, Pl. 1:6. □1978 *Eliasum llaniscum* Fombella 1977 – Fombella, Pl. 1:3. □1979 *Cymatiosphaera* sp. 1 – Volkova *et al.*, p. 26, Pl. 16:6–8. □1981 *Eliasum llaniscum* Fombella, 1977 – Martin & Dean, p. 19, Pl. 2:14. □1981 *Eliasum llaniscum* Fombella – Erkmen & Bozdoğan, p. 54, Pl. 2:15–16. □1983 *Eliasum llaniscum* Fombella, 1977 – Vanguestaine & Van Looy, Pls. 1:9, 11; 2:17. □1983 *Eliasum llaniscum* Fombella, 1977 – Martin & Dean, Pl. 43.2, Fig. 8. □1984 *Eliasum llaniscum* Fombella, 1978 – Martin & Dean, Pl. 57.2, Figs. 6, 8, 10–15. □1986 *Eliasum* cf. *E. llaniscum* – Welsch, p. 66, Pl. 2:9, 10. □1988 *Eliasum llaniscum* Fombella, 1977 – Martin & Dean, Pl. 15:8, 9. □1988 *Eliasum* cf. *E. llaniscum* Fombella 1977 – Bagnoli, Stauge and Tongiorgi, Pls. 26:1–5; 27:1–5. □1989 *Eliasum llaniscum* Fombella, 1977 – Mette, p. 2, Pl. 1:2. □1989 *Eliasum llaniscum* – Hagenfeldt 1989a, pp. 46–48, Pl. 2:9. □1989 *Eliasum llaniscum* – Fatka, p. 365, Pl. 1:7. □1990 *Eliasum llaniscum* – Volkova, Pl. 1:3, 6, 7. □1990 *Eliasum* sp. – Eklund, Fig. 10B. □1993 *Eliasum llaniscum* Fombella, 1977 – Fombella *et al.* Pl. 2:1. □1994 *Eliasum llaniscum* Fombella, 1977 – Martin *in* Young *et al.* Fig. 10h, k. □1995 *Eliasum llaniscum* – Moczydłowska & Crimes, p. 120–121, Pl. 1A, B.

Material. – One-hundred-and-three well-preserved specimens.

Description. – Thin-walled, elongated oval vesicles with smooth surface and rounded poles having few (two or three) wide crests extending along the vesicle. The crests are approximately of equal width, parallel in the central part of the vesicle and becoming narrower and closer at the polar parts of the vesicle.

Dimensions. – $N=22$. Vesicle width 25–40 μm, vesicle length 60–110 μm, width of crests 5–9 μm.

Remarks. – The micropunctate ornamentation on the vesicle wall between the crests was observed under the light microscope on specimens referred to *Eliasum* cf. *E. llaniscum* Fombella, 1977, by Welsch (1986, p. 66, Pl. 2:9–10). These specimens do not otherwise differ from *E. llaniscum*, diagnosed by Fombella (1977) as having smooth surface of the vesicle and crests. Such fine ornamentation, if present, is usually not observable in light microscopy and is easily obscured by the state of preservation of microfossils. Thus, it has no practical significance for the taxonomy of this species.

The ultrastructure of the vesicle wall (irregularly micro-corrugated) and the crests (microgranulate) was revealed

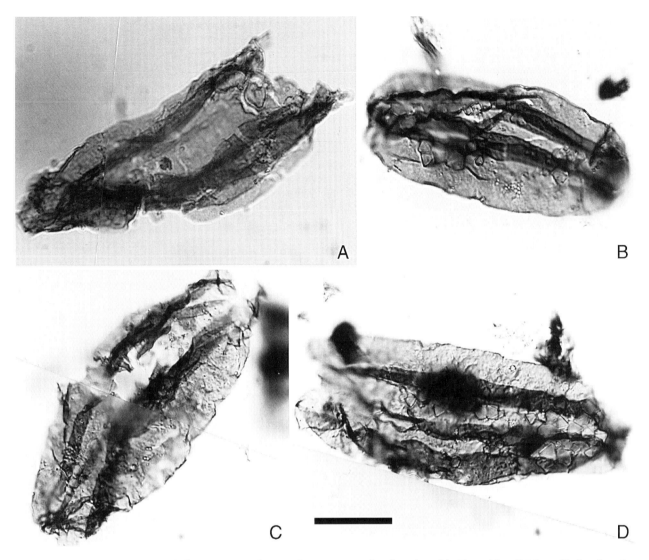

Fig. 28. □A–D. *Eliasum llaniscum* Fombella. Specimens with two or three crests extending along the vesicle. The vesicle wall deformed by imprints after the pyrite pseudomorphs is shown in B. and D. A, PMU-Pl.54-U/46; B, PMU-Pl.55-U/48/3; C, PMU-Pl.56-K/34; D, PMU-Pl.57-G/35/3. All specimens from the Sosnowiec IG-1 borehole, Sosnowiec formation. A, depth 3403.5–3407.2 m, Middle Cambrian, *Acadoparadoxides oelandicus* Superzone; B, C, D, depth 3174.0–3174.7 m, lower Upper Cambrian. Scale bar in D equals 15 µm for A; 18 µm for B; 20 µm for C, D.

under the high magnification in SEM on specimens of *Eliasum* cf. *E. llaniscum* Fombella, 1977, by Bagnoli *et al.* (1988, pp. 194–195, Pl. 26:4; 27:5). It was rarely observed among other specimens with a psilate surface and has no value as a potential taxonomic feature. The microsculpture on the wall and crests shown on the light-microscope photomicrographs (Bagnoli *et al.* 1988, Pls. 26:1, 3, 5; 27:1, 3) looks like the effect of poor state of preservation of the microfossils, in which vesicles are degraded by the imprints of the pyrite crystals and by having fragments of particulate organic matter superimposed on the vesicle.

Present record. – The Upper Silesia area, Sosnowiec IG-1 borehole, Sosnowiec formation in the interval of 3174.0–

3423.8 m, Middle Cambrian *Acadoparadoxides oelandicus* Superzone to lower Upper Cambrian.

Occurrence and stratigraphic range. – The species has wide geographic distribution, being commonly recorded in the Middle Cambrian *Acadoparadoxides oelandicus* and *Paradoxides paradoxissimus* Superzones (see compilation below). Vanguestaine & Van Looy (1983, p. 78) and Downie (1984, p. 9) indicated the range of the species extending into the middle part of the *Paradoxides forchhammeri* Superzone or throughout the zone, respectively. However, the evidence for establishing this portion of the range (geologic successions and fossil records) was not provided. In the present study, the stratigraphic range of the species is inferred to extend into the *P. forchhammeri*

Superzone and also into the Upper Cambrian. This conclusion is entirely based on micropalaeontological data. In consistency herewith, the record of the species in Sierra Morena in Spain is suggested to be Late Cambrian in age (see under 'Stratigraphic ranges' and below).

The occurrence of the species in the uppermost Lower Cambrian was reported from Latvia in the Rausve 'horizon' by Volkova *et al.* (1979), but the stratigraphic position of the strata concerned has never been supported by macrofossils, nor has the stratigraphic succession been published. The species was recorded in Sweden, Gotska Sandön Island, in the uppermost File Haidar Formation attributed to the Lower Cambrian *Proampyx linnarssoni* Zone (Hagenfeldt 1989a). This stratigraphic attribution is uncertain and can be alternatively considered to be Middle Cambrian (Eklund 1990; Moczydłowska 1991; see under 'Re-evaluation of stratigraphic ranges'). In NW Wales the species occurs in St. Tudwal's Peninsula and St. Tudwal's Island East in the Hell's Mouth Formation, referred to the Lower Cambrian Protolenid–Strenuellid Zone (Young *et al.* 1994). However, neither trilobite taxa nor the acritarch assemblage in this formation are sufficiently diagnostic to recognize the Lower–Middle Cambrian boundary in the succession (see under 'Re-evaluation of stratigraphic ranges').

Detailed record of occurrences: Spain, the Cantabrian Mountains at Láncara de Luna, the Oville Formation, upper Middle Cambrian (Cramer & Diez 1972); the Oville Formation referred to as Middle Cambrian to Lower Tremadocian by Fombella (1977, 1978, 1979), but the stratigraphic range of this formation is documented only as Middle Cambrian (see under 'Stratigraphic ranges'); Sierra Morena, Province of Huelva, Cañaveral de Leon, Umbria Pipeta Formation, Middle Cambrian (Mette 1989; the relative age of this formation is considered here as probably late Cambrian, see under 'Stratigraphic ranges'). Canada, Newfoundland, Random Island and Manuels River, Chamberlains Brook Formation and Manuels River Formation, Middle Cambrian, *Acadoparadoxides oelandicus* and *P. paradoxissimus* Superzones (Martin & Dean 1981, 1983, 1984, 1988). Latvia, Vergale-49 borehole (1238.0 m) and Liepaya borehole (1363.6 m), Middle Cambrian, Kibartai horizon (=*Acadoparadoxides oelandicus* Superzone) (Volkova *et al.* 1979; Volkova 1990). Sweden, Östergötland, upper part of the File Haidar Formation and 'oelandicus mudstones', Middle Cambrian, *Acadoparadoxides oelandicus* Superzone (Eklund 1990); south-central area, Närke at Kvantorp, Island of Gotland and Gotska Sandön Island, 'oelandicus beds', Gulf of Bothnia, Finngrundet borehole, upper Söderfjärden Formation, Middle Cambrian, *Acadoparadoxides oelandicus* Superzone (Hagenfeldt 1989a, b); Island of Öland, Furuhäll near Borgholm, Djupvik formation, Middle Cambrian, *Paradoxides paradoxissimus* Superzone (Bagnoli *et al.* 1988). Finland, Vaasa area,

Söderfjärden 3 borehole, upper Söderfjärden Formation, Middle Cambrian, *Acadoparadoxides oelandicus* Superzone (Hagenfeldt 1989b). SE Turkey, Merdin–Derik area, Sosink Formation – Middle Cambrian (Erkmen & Bozdoğan 1981). Morocco, High Atlas Mountains, Tacheddirt Valley, Middle Cambrian, *Acadoparadoxides oelandicus* and *P. paradoxissimus* Superzones (Vanguestaine & Van Looy 1983). Norway, Finnmark, Digermul Peninsula, Kistedal Formation, Middle Cambrian, *Paradoxides paradoxissimus* Zone (Welsch 1986). Czech Republic, Bohemian Massif, Jince, Jince Formation, Middle Cambrian, *Eccaparadoxides pusillus* Zone (Vavrdová 1976), and Vinice Hill near Jince, Jince Formation, Middle Cambrian, *Onymagnostus hybridus* Zone (=*Tomagnostus fissus* and *Ptychagnostus atavus* Zone) (Fatka 1989). NW Wales, St. Tudwal's Peninsula and St. Tudwal's Island East, Hell's Mouth Formation, Lower Cambrian, *Protolenid–Strenuellid* Zone; Trwyn y Fulfran Formation and Cilan Formation, Middle Cambrian, *Acadoparadoxides oelandicus* Superzone; Ceriad Formation and Nant-y-big Formation, Middle Cambrian, within the range of *A. oelandicus* and *Paradoxides paradoxissimus* Superzone (Young *et al.* 1994). Ireland, Co. Wexford, the Booley Bay Formation, upper part of Upper Cambrian inferred from the acritarch record (Moczydłowska & Crimes 1995).

Genus *Estiastra* Eisenack, 1959

Type species. – Estiastra magna Eisenack, 1959, pp. 201–202, Pl. 16:17–20; Estonia, lowermost Silurian.

Estiastra minima Volkova, 1969
Fig. 27D–E

Remarks. – The emendation to the diagnosis of the genus *Estiastra* by Sarjeant & Stancliffe (1994, p. 50) is not adopted here, because it did not provide any significant morphologic details that would lead to a better understanding of the original diagnosis. The proposed limitation of the number of processes to 4–10 is unjustified and is inconsistent with observations on some species having more processes (*E. improcera* in Loeblich 1970, p. 720; *E. minima* in Volkova *et al.* 1979, Pl. 10:10). Furthermore, the proposed emendation lacks any statement concerning the relationship between the cavity of vesicle and processes, one of the major diagnostic features.

Synonymy. – □1969 *Estiastra minima* Volkova sp.nov. – Volkova, 1969b, pp. 230–321, Pl. 50:32–36. □1979 *Estiastra minima* Volkova, 1969 – Volkova *et al.*, p. 18, Pl. 10:8–14. □1982 *Veryhachium* sp. – Wood & Clendening, p. 263, Pl. 2:12. □1987 *Estiastra minima* – Knoll & Swett, p. 916, Fig. 8.13. □1989 *Estiastra minima* – Hagenfeldt,

1989a, Pl. 2:10. □1991 *Estiastra minima* – Moczydłowska 1991, p. 53, Pl. 9A–G. □1994 *Estiastra? minima* Volkova 1969 – Sarjeant & Stancliffe, p. 51.

Material. – Five well-preserved specimens.

Description. – Vesicle polygonal in outline, having an irregular central body and bearing 5–10 large conical processes with wide bases and sharp-pointed tips. The cavity of the central body and processes communicate freely.

Dimensions. – N=5. Diameter of central body 7–18 μm, length of processes 3–9 μm. Width of basal part of processes is 3–7 μm.

Remarks. – The nature of the connection between the cavities of the processes and the central body was not mentioned in the diagnosis of the species (Volkova 1969b). The holotype and specimens from the type collection show an opaque central body, owing to their state of preservation, and transparent processes (Volkova 1969b, Pl. 50:32–36) that appear not to communicate with the vesicle cavity. Moczydłowska (1991) assumed that such a connection was missing in *E. minima*. Observations on very well-preserved specimens from the Baltic area have shown that the vesicle interior communicates with the cavities of the processes (Hagenfeldt 1989a, p. 48, Pl. 2:10).

Present record. – The Upper Silesia area, Sosnowiec IG-1 borehole, Sosnowiec formation at depths of 3403.5–3407.2 m and 3365.2–3372.3 m; Middle Cambrian, *Acadoparadoxides oelandicus* Superzone.

Occurrence and stratigraphic range. – Poland, East European Platform, Lublin Slope, Kaplonosy and Radzyń formations, and Podlasie Depression, 'Holmia zone s.l.', Lower Cambrian, *Holmia kjerulfi* and *Protolenus* Zones (Volkova 1969b; Moczydłowska 1991). Latvia, Lower Cambrian, Vergale and Rausve horizons, and Middle Cambrian, Kibartai horizon; Lithuania and the Ukraine, Lower Cambrian, Vergale horizon (Volkova *et al.* 1979). U.S.A, Tennessee, Chilhowee Group, Murray Shale, Lower Cambrian (Wood & Clendening 1982). Svalbard, East Spitsbergen: South Tokammane, Tokammane Formation, Lower Cambrian (Knoll & Swett 1987). Sweden, Östergötland, Island of Gotland and Gotska Sandön Island, File Haidar Formation and Gulf of Bothnia, Söderfjärden Formation, Lower Cambrian, *Holmia kjerulfi* and *Proampyx linnarssoni* Zones; Gotska Sandön Island, 'oelandicus beds' – Middle Cambrian, *Acadoparadoxides oelandicus* Superzone (Hagenfeldt 1989a; Eklund 1990).

Genus *Globosphaeridium* Moczydłowska, 1991

Type species. – *Globosphaeridium cerinum* (Volkova, 1968) Moczydłowska, 1991, p. 54 (=*Baltisphaeridium cerinum* Volkova, 1968, pp. 17–18, Pl. 1:1 (holotype) and Pls. 1:2–7; 11:5); Estonia, Pirita Formation, Lower Cambrian, Talsy horizon (time equivalent to *Schmidtiellus mickwitzi* Zone).

Remarks. – The proposed transfer of the genus *Globosphaeridium* Moczydłowska, 1991 to *Filisphaeridium* Staplin *et al.*, 1965, as its junior synonym by Sarjeant & Stancliffe (1994, p. 6) is groundless because of the significant differences in the morphology of the processes in both genera. *Globosphaeridium* has processes tapering towards distal portions with simple tips (Moczydłowska 1991, p. 54), contrary to the cylindrical and wiry processes with unbranched or distally differentiated tips (thickened or branching) of *Filisphaeridium* (Staplin *et al.* 1965, p. 192). The type species of *Filisphaeridium* (*F. setasessitante*) has clearly capitate process tips (Staplin *et al.* 1965). The morphology of the process terminations used to define the above genera, both having solid processes, parallels that distinguishing *Baltisphaeridium* (acuminate tips) from *Skiagia* (funnel-like tips), two genera having hollow processes separated from the vesicle cavity (Eisenack 1958b, 1969; Downie 1982; Moczydłowska 1991).

Globosphaeridium cerinum (Volkova, 1968) Moczydłowska, 1991

Synonymy and description. – See Moczydłowska 1991, p. 54.

Material. – Six poorly preserved specimens.

Dimensions. – N=6. Vesicle diameter 24–30 μm, process length 3–4 μm.

Present record. – The Upper Silesia area, Goczałkowice IG-1 borehole, Goczałkowice formation at a depth of 2969.2–2973.7 m, Lower Cambrian, *Holmia kjerulfi* Zone.

Occurrence and stratigraphic range. – The species has been recorded in Estonia, Latvia, Poland, the Ukraine, Russia, Kazakhstan, Sweden, Denmark, Norway, Belgium, East and North Greenland in the Lower Cambrian, the *Schmidtiellus mickwitzi* and *Holmia kjerulfi* Zones and their time equivalents (see compilation by Moczydłowska 1991, pp. 54–55). An additional occurrence is in the Buen Formation, Peary Land of North Greenland, within the same stratigraphic range (Vidal & Peel 1993).

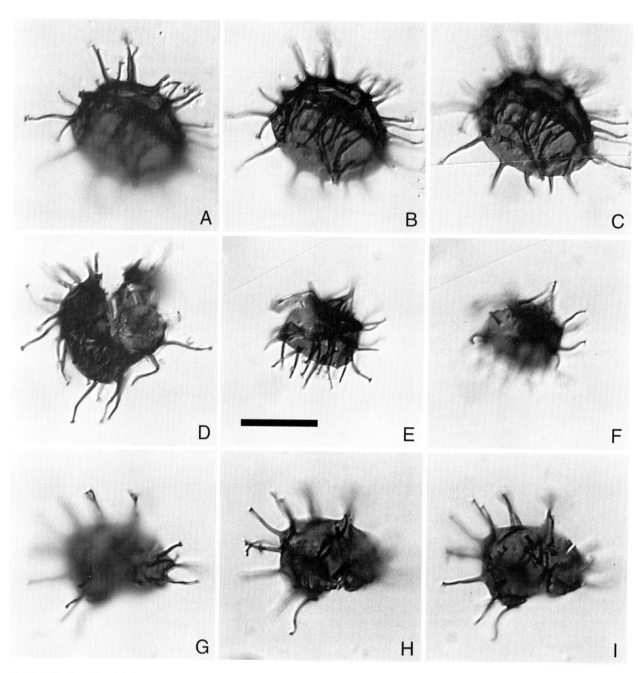

Fig. 30. Heliosphaeridium bellulum n.sp. □A–C. PMU-Pl.63-E/33/1, holotype, at various focus levels showing three-dimensional preservation of the vesicle, with smooth wall surface, and processes. The attachment of processes is clearly visible in B and C. The hollow process cavities are freely connected with the vesicle interior, as shown in C. □D. PMU-Pl.64-N/44, paratype, displaying diagnostic regular outline of the vesicle and straight attachment of processes. The processes are tubular in their major portion and may slightly taper at the distal end, terminated by distinctive button-like tips, shown particularily well here, and in E and H. □E–F. PMU-Pl.65-U/30, at two focus levels; E, at high focus with processes of equal length; and F, low focus. □G–I. PMU-Pl.66-S/31/4, at three focus levels showing very regular morphology of processes. The Sosnowiec IG-1 borehole, depth 3403.5–3407.2 m, Sosnowiec formation, Middle Cambrian, *Acadoparadoxides oelandicus* Superzone. Scale bar in E equals 12 μm for all micrographs.

łowska, 1991, in general morphology and in the presence of funnel-like tips on the processes. *H. notatum* differs from the present species by having wide process bases that result in a more wavy outline on the vesicle, and by having wider and more conical processes.

Present record. – The Upper Silesia area, Sosnowiec IG-1 borehole, Sosnowiec formation at depths of 3353.3–3359.3 m and 3403.5–3407.2 m, Middle Cambrian, *Acadoparadoxides oelandicus* Superzone.

Heliosphaeridium coniferum (Downie, 1982) Moczydłowska, 1991

Fig. 31A–C

Synonymy. – □1982 *Micrhystridium coniferum* sp.nov. – Downie, p. 260, Fig. 6q–t. □*non* 1983 *Micrhystridium* aff. *coniferum* Downie 1982 – Vanguestaine & Van Looy, p. 73, Pl. 1:16–19. □1987 *Micrhystridium* cf. *M. minutum* Downie, 1982 – Knoll & Swett, p. 919, Fig. 8.3. □1991 *Heliosphaeridium coniferum* (Downie, 1982) comb.nov. – Moczydłowska, pp. 58–59, Pl. 8B–D. □1993 *Heliosphaeridium coniferum* (Downie, 1982) Moczydłowska, 1991 – Vidal & Peel, p. 25, Fig. 8g, k. □1994 *Micrhystridium coniferum* Downie 1982 – Sarjeant & Stancliffe, p. 16.

Material. – Four well-preserved specimens.

Description. – Minute vesicles, oval in outline, originally spherical, bearing numerous processes that are evenly and densely distributed. The processes are of equal length, having wide conical bases that taper towards slim distal portions with blunt or truncated tips. The bases of the processes are hollow and communicate with the inner cavity of the vesicle.

Dimensions. – $N=4$. Diameter of central body 5–7 µm, process length 2–4 µm.

Present record. – The Upper Silesia area, Sosnowiec IG-1 borehole, Sosnowiec formation at a depth of 3365.2–3372.3 m, Middle Cambrian, *Acadoparadoxides oelandicus* Superzone.

Occurrence and stratigraphic range. – Scotland, Fucoid Beds, Lower Cambrian (Downie 1982). Canada, Alberta, Gog Formation, Lower Cambrian (Downie 1982). Svalbard, East Spitsbergen, South Tokammane and Tokammane Formation, Lower Cambrian (Knoll & Swett 1987). Poland, the East European Platform, Kaplonosy and Radzyń formations, Lower Cambrian, *Schmidtiellus mickwitzi* and *Holmia kjerulfi* Zones (Moczydłowska 1991). Greenland, Peary Land, Buen Formation, Lower Cambrian, equivalent to *Holmia kjerulfi* – *Protolenus* Zones (Vidal & Peel 1993).

Heliosphaeridium dissimilare (Volkova, 1969) Moczydłowska, 1991

Fig. 32H

Synonymy. – See Moczydłowska 1991, p. 58. Additionally: □1986 *Micrhystridium dissimilare* Volkova 1969 – Welsch, pp. 53–54, Pl. 1:10. □1989 *Micrhystridium dissimilare* Volkova 1969 – Hagenfeldt 1989a, pp. 72–74, Pl. 3:11. □1993 *Heliosphaeridium dissimilare* (Volkova, 1969) Moczydłowska, 1991 – Vidal & Peel, p. 25, Fig. 9d. 1994 *Comasphaeridium dissimilare* (Volkova 1969, pp.

262–263, pl. 2, figs 12–13, 19–29) comb.nov. – Sarjeant & Stancliffe, p. 26. (Note erroneously cited reference; it should be: Volkova 1969b, p. 227, Pl. 50:12–13, 19–20). □1995 *Heliosphaeridium dissimilare* (Volkova 1969) Moczydłowska 1991 – Vidal *et al.*, Fig. 5:3, 4, 8.

Material. – Twelve well-preserved specimens.

Description. – See Moczydłowska 1991, p. 58.

Dimensions. – $N=10$. Diameter of central body 7–14 µm, process length 4–5 µm.

Remarks. – The species has been reported to occur in Russia, in the St. Petersburg area, in the Izhora beds which were originally referred to the Lower? Cambrian (Volkova *et al.* 1979). The relative age of these strata has been revised subsequently and attributed to the Late Cambrian (Jankauskas 1980; Volkova 1990; Moczydłowska & Crimes 1995). However, the occurrence of *H. dissimilare* in the Izhora beds has never been illustrated nor subsequently (i.e. after the year 1979) confirmed. It is regarded here as uncertain. *H. dissimilare* was erroneously reported in the Radzyń IG-1 borehole in Poland, in rocks attributed to the *Platysolenites* Zone and *Paradoxides paradoxissimus* Zone (Moczydłowska 1991, p. 58); this record should be obliterated.

Present record. – The Upper Silesia area, Goczałkowice IG-1 borehole, Goczałkowice formation at a depth of 2766.8–2771.1 m, and Sosnowiec IG-1 borehole, Sosnowiec formation in the interval of 3353.3–3407.2 m, Middle Cambrian, *Acadoparadoxides oelandicus* Superzone.

Occurrence and stratigraphic range. – The species has been recorded in Poland, Lithuania, Latvia, Estonia, the Ukraine, Sweden, Norway, Scotland, Svalbard, Greenland, Canada and Kazakhstan in the Lower Cambrian, time equivalent to *Holmia kjerulfi* and *Protolenus* Zones, and in the Middle Cambrian, strata of the *Acadoparadoxides oelandicus* Superzone (detailed summary by Moczydłowska 1991). Additionally in Sweden, Island of Gotland, Grötlingbo-1 borehole, Gotska Sandön Island, Gotska Sandön borehole, Närke at Kvarntorp, 'oelandicus beds', the Gulf of Bothnia, Finngrundet borehole, Söderfjärden Formation, Middle Cambrian, *Acadoparadoxides oelandicus* Superzone (Hagenfeldt 1989a). Finland, Vaasa area, Söderfjärden Formation, Middle Cambrian, *Acadoparadoxides oelandicus* Superzone (Hagenfeldt 1989a). Norway, Finnmark, Digermul Peninsula, Kistedal Formation, Middle Cambrian, *Paradoxides paradoxissimus* Zone (Welsch 1986). Spain, the Iberian Mountains, Ribota Formation and Daroca Sandstones, Lower Cambrian (Gamez *et al.* 1991), time equivalent to the *Holmia kjerulfi* and *Protolenus* Zones, respectively; the Cantabrian Mountains at Irede de Luna, Herreria Formation, Lower Cambrian, *Holmia kjerulfi* Zone (Palacios & Vidal 1992). Greenland, Peary Land, Buen Formation, Lower Cambrian, *Holmia*

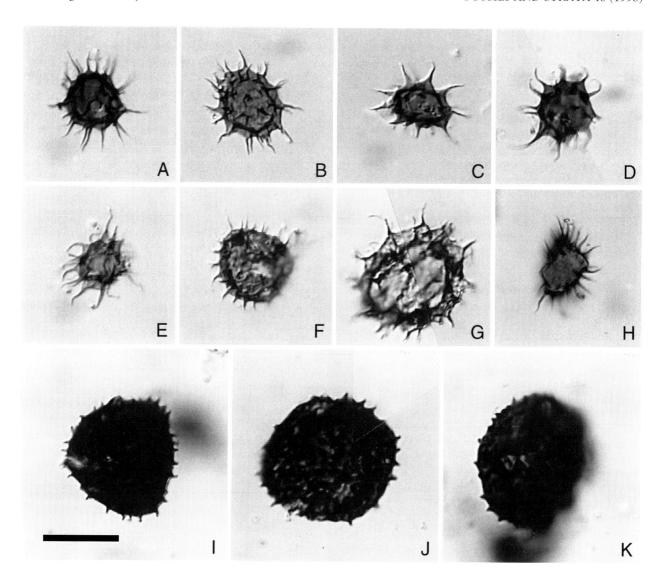

Fig. 32. □A–E. *Heliosphaeridium lubomlense* (Kirjanov) Moczydłowska. A, PMU-Pl.82-H/40/4; B, PMU-Pl.83-S/35; C, PMU-Pl.84-X/42; D, PMU-Pl.85-T/29/1; E, PMU-Pl.86-P/43/4. Specimens with diagnostic widened conical bases and tapering distally processes. □F. *Heliosphaeridium nodosum* n.sp. PMU-Pl.87-U/29; holotype, showing rigid thorn-like short processes. □G. *Heliosphaeridium exile* n.sp. PMU-Pl.88-V/41; holotype, with diagnostic slender and solid distal portions of the processes and conical hollow process bases, forming concave outline of vesicle wall between of them. □H. *Heliosphaeridium dissimilare* (Volkova) Moczydłowska. PMU-Pl.89-E/40/1; showing processes gradually tapering towards the tips from elongated conical bases. □I–K. *Heliosphaeridium serridentatum* n.sp. I, PMU-Pl.90-X/38/3; holotype, displaying numerous short and rigid processes forming sawtoothed outline of the vesicle. J, PMU-Pl.91-T/40/2; K, PMU-Pl.92-Y/43/4. All specimens from the Sosnowiec IG-1 borehole, Sosnowiec formation. A–E, H, depth 3403.5–3407.2 m, and F, I, J, K, depth 3365.2–3372.3 m, Middle Cambrian, *Acadoparadoxides oelandicus* Superzone; G, depth 3166.3–3167.7 m, lower Upper Cambrian. Scale bar in I equals 12 μm for all micrographs.

Lower Tremadoc by Fombella (1986), but the stratigraphic range of the formation in this area spans only Middle Cambrian (Aramburu *et al.* 1992, see under 'Stratigraphic ranges'). Russia, Kaliningrad area, Veselovsk 5 borehole, Veselovsk Formation, Middle Cambrian, upper *Paradoxides paradoxissimus* to lower *P. forchhammeri* Zone (Volkova 1990).

Heliosphaeridium longum (Moczydłowska, 1988) Moczydłowska, 1991

Fig. 31G–H

Synonymy. – □1988 *Micrhystridium longum* n.sp. – Moczydłowska 1988, pp. 7–8, Pl. 1:5, 6; Fig. 3. □1989 *Mic-*

rhystridium sp. 1 – Hagenfeldt 1989b, pp. 210–211, Pl. 2:10. □1991 *Heliosphaeridium longum* (Moczydłowska, 1988) comb.nov. – Moczydłowska, p. 59, Pl. 8P–Q. □1994 *Micrhystridium longum* Moczydłowska 1988 – Sarjeant & Stancliffe, p. 17.

Material. – Seven poorly preserved specimens.

Description. – Vesicles circular to oval in outline, originally spherical, bearing long and slender processes, approximately 7–15 observed on the surface. The processes have elongated conical bases and gradually taper in the distal portions. The terminations of processes are acute. The processes are hollow and connected with the vesicle cavity.

Dimensions. – $N=5$. vesicle diameter 6–14 µm, process length 4–9 µm.

Present record. – The Upper Silesia area, Sosnowiec IG-1 borehole, Sosnowiec formation at depths of 3210.0–3212.0 m and 3174.0–3174.7 m; Middle Cambrian, *Paradoxides paradoxissimus* Superzone and lower Upper Cambrian, respectively.

Occurrence and stratigraphic range. – Poland, Lublin Slope of the East European Platform, Kaplonosy and Radzyń formations, Lower Cambrian, *Protolenus* Zone (Moczydłowska 1991). Sweden, Gotska Sandön Island, 'oelandicus beds', Middle Cambrian, *Acadoparadoxides oelandicus* Superzone (Hagenfeldt 1989b).

Heliosphaeridium lubomlense (Kirjanov, 1974) Moczydłowska, 1991

Fig. 32A–E

Synonymy. – See Moczydłowska 1991, p. 59. Additional: □1986 *Micrhystridium lubomlense* Kirjanov 1974 – Welsch, p. 55, Pl. 1:11–14. □1993 *Heliosphaeridium lubomlense* (Kirjanov, 1974) Moczydłowska, 1991 – Vidal & Peel, p. 25, Fig. 8a, b. □1994 *Micrhystridium? lubomlense* Kirjanov 1974 – Sarjeant & Stancliffe, p. 17.

Material. – Twenty-eight well-preserved specimens.

Description. – See Moczydłowska 1991, p. 59.

Dimensions. – $N=10$. Vesicle diameter 7–18 µm, process length 4–7 µm.

Present record. – The Upper Silesia area, Goczałkowice IG-1 borehole, Goczałkowice formation at a depth of 2766.8–2771.1 m, and Sosnowiec IG-1 borehole, Sosnowiec formation in the interval of 3353.3–3407.2 m, Middle Cambrian, *Acadoparadoxides oelandicus* Superzone.

Occurrence and stratigraphic range. – The species has been recorded in the Ukraine, Latvia, Lithuania, Kazakhstan,

Svalbard and Poland in the Lower Cambrian, time equivalent to *Holmia kjerulfi* and *Protolenus* Zones, and in the Middle Cambrian, *Acadoparadoxides oelandicus* Superzone (see compilation by Moczydłowska 1991). Additional records are from Norway, Finnmark, Digermul Peninsula, Kistedal Formation, Middle Cambrian, *Paradoxides paradoxissimus* Zone (Welsch 1986), and Greenland, Peary Land, Buen Formation, Lower Cambrian, part of the *Holmia kjerulfi* and *Protolenus* Zones (Vidal & Peel 1993).

Heliosphaeridium nodosum n.sp.

Fig. 32F

Holotype. – Specimen PMU-Pl.87-U/29; Fig. 32F.

Derivation of name. – From Latin *nodosus* – thorny, referring to the shape of the processes.

Locus typicus. – The Upper Silesia area, Sosnowiec IG-1 borehole (Figs. 2 and 3).

Stratum typicum. – Alternating mudstones and sandstones in the Sosnowiec formation at a depth of 3365.2–3372.3 m; Middle Cambrian, *Acadoparadoxides oelandicus* Superzone.

Material. – Eight well-preserved specimens.

Diagnosis. – Vesicles circular to oval in outline, originally spherical, bearing thorn-like processes, rigid in appearance. Processes, approximately 20–25 observed on the vesicle contour, have widened basal portions gradually tapering towards sharp-pointed tips. They are hollow and are connected with the cavity of the vesicle.

Dimensions. – $N=8$. Vesicle diameter 5–13 µm, process length 2–4 µm.

Remarks. – The species differs from *H. obscurum* in having more numerous (approximately twice as many as observed on the vesicle contour) and evenly distributed processes which gradually taper towards the tips and have less developed bases.

Present record. – The Upper Silesia area, Sosnowiec IG-1 borehole, Sosnowiec formation at depths of 3365.2–3372.3 m and 3210.0–3212.0 m; Middle Cambrian, *Acadoparadoxides oelandicus* and *Paradoxides paradoxissimus* Superzones, respectively.

Heliosphaeridium notatum (Volkova, 1969) Moczydłowska, 1991

Figs. 31L, 33A–L

Synonymy. – □1969 *Micrhystridium notatum* Volkova sp.nov. – Volkova 1969b, p. 228, Pl. 51:16–19. □1974

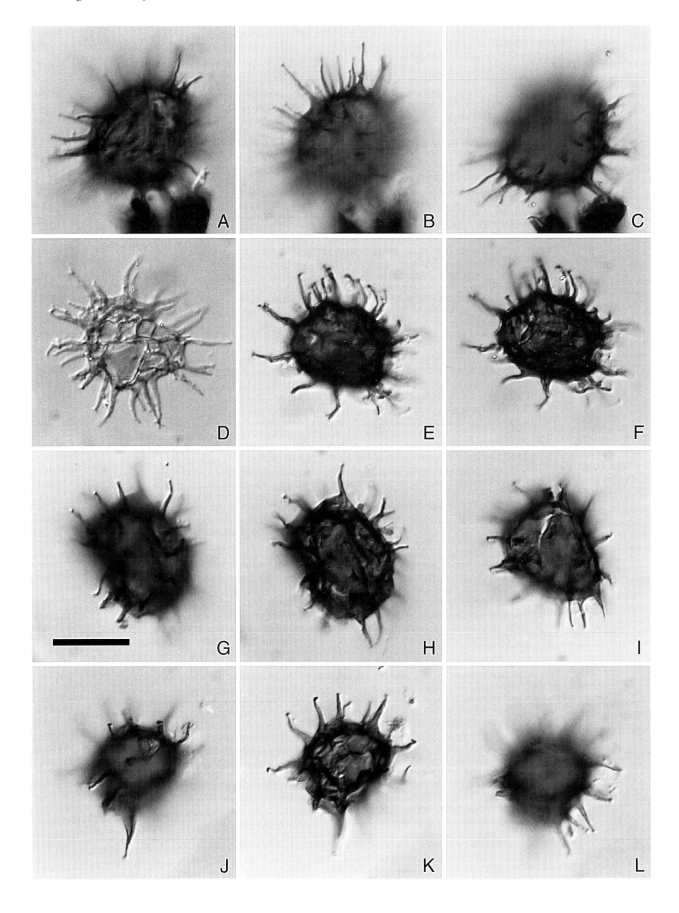

Micrhystridium notatum – Volkova, Pl. 17:10–11. □1979
Micrhystridium notatum – Volkova *et al.*, p. 15, Pl. 9:1–4.
□1983 *Micrhystridium* aff. *coniferum* Downie 1982 –
Vanguestaine & Van Looy, p. 73, Pl. 1:16–19. □1986 *Micrhystridium notatum* Volkova 1979 – Fombella, Pl. 2:12.
□1989 *Micrhystridium notatum* – Hagenfeldt 1989a, Pl.
4:1. □1991 *Heliosphaeridium notatum* (Volkova, 1969)
comb.nov. – Moczydłowska, p. 58. □1992 *Heliosphaeridium notatum* (Volkova) Moczydłowska, 1991 – Palacios &
Vidal, Fig. 6d. □1994 *Micrhystridium notatum* Volkova
1969 – Sarjeant & Stancliffe, p. 17.

Material. – One-hundred-and-twelve well-preserved
specimens.

Description. – Small vesicles circular to oval in outline,
originally spherical, bearing numerous (approximately
15–25 observed on the vesicle outline) and evenly distributed processes whose length is approximately equal to the
radius of the vesicle. Processes are tubular or taper
towards the terminations and have distinctive conical
bases. The tips of the processes are in the shape of small
closed funnels; occasionally some tips are simply truncated or might be bifurcated. The conical bases of the
processes arise very gradually from the wall of the vesicle
and give the vesicle an undulating outline. The processes
are hollow, and their cavity is connected freely with the
inner cavity of the vesicle.

Dimensions. – $N=37$. Diameter of the vesicle is 9–23 μm,
length of processes 4–9 μm. One exceptionally small specimen has a diameter of 5×11 μm and length of processes
3–4 μm.

Remarks. – The length of the processes is almost constant,
with very little variation among specimens. Within the
population of *H. notatum* (within one sample), there
occur specimens that display variably shaped process terminations. A few processes among the bulk of processes
having funnel-like tips might have dichotomizing ends.
There is a full array of specimens with an increasing

Fig. 33. □A–L. *Heliosphaeridium notatum* (Volkova) Moczydłowska.
Intraspecific variation of the species. A–C, PMU-Pl.93-Z/26/4, three-dimensionally preserved specimen at various focus levels, showing gradually tapering processes with conspicous conical basal portions; D,
PMU-Pl.94-E/31/2, some process tips are bifurcated, as visible in upper
left quadrant of the specimen. □E–F. PMU-Pl.95-I/32, two focus levels
showing regular morphology of processes. □G–H. PMU-Pl.96-V/32,
different focus levels, well-preserved funnel-like tips of processes. □I.
PMU-Pl.97-P/28; some processes might be simple tubular with truncated tips, as seen in the lower part of the specimen. □J–L. PMU-Pl.98-V/32/2; various focus levels showing a single process conical in shape
with truncated tip among the majority of processes with funnel terminations. All specimens from the Sosnowiec IG-1 borehole, Sosnowiec Formation. A–C, E–L, depth 3403.5–3407.2 m, and D, depth 3365.2–3372.3
m, Middle Cambrian, *Acadoparadoxides oelandicus* Superzone. Scale bar
in G equals 12 μm for all micrographs.

number of processes that have bifurcated terminations.
This comprises 'typical' *H. notatum* with homomorphic
processes with funnel tips, specimens with a single bifurcated process, and specimens with a few bifurcated processes. The latter morphotype resembles, and might in fact
be a transitional form to, the genus *Multiplicisphaeridium*.

Present record. – The Upper Silesia area, Sosnowiec IG-1
borehole, Sosnowiec formation at depths of 3419.6–
3423.8 m, 3403.5–3407.2 m and 3365.2–3372.3 m, Middle
Cambrian, *Acadoparadoxides oelandicus* Superzone; and
at a depth of 3210.0–3212.0 m, Middle Cambrian, *Paradoxides paradoxissimus* Superzone;

Occurrence and stratigraphic range. – Poland, Lublin
Slope of the East European Platform, borehole Radzyń
IG-1, Radzyń formation, upper Lower Cambrian (*Protolenus* Zone), and Kostrzyń Formation, Middle Cambrian,
Acadoparadoxides oelandicus Zone (Volkova 1969b, Volkova *et al.* 1979). Latvia, Lower Cambrian, Rausve horizon;
Latvia, Lithuania and Russia, Middle Cambrian, Kibartai
horizon; Lithuania, Middle Cambrian, Deimenos Group
(Volkova *et al.* 1979). Sweden, Gotska Sandön Island,
Gotska Sandön borehole, upper File Haidar Formation,
Lower Cambrian, *Proampyx linnarssoni* Zone (Hagenfeldt
1989a; age of the strata uncertain, might be middle Cambrian – see comments under 'Stratigraphic ranges');
Östergötland, File Haidar Formation, Middle Cambrian
(Eklund 1990); Gulf of Bothnia, Finngrundet borehole,
Söderfjärden Formation, Middle Cambrian; Gotland
Island, Grötlingbo-1 borehhole, Gotska Sandön Island,
Gotska Sandön borehole, and south-central Sweden,
Närke at Kvarntorp, ´oelandicus beds', Middle Cambrian,
Acadoparadoxides oelandicus Stage (Hagenfeldt 1989a).
Finland, Vaasa area, Söderfjärden-3 borehole, Söderfjärden Formation, Middle Cambrian *Acadoparadoxides
oelandicus* Stage (Hagenfeldt 1989a). Morocco, High
Atlas Mountains, Tacheddirt Valey, lower Middle Cambrian (Vanguestaine & Van Looy 1983). Spain, the Cantabrian Mountains at Irede de Luna, Herreria Formation,
Lower Cambrian, equivalent to *Protolenus* Zone (Palacios
& Vidal 1992). North Greenland, Peary Land, upper Buen
Formation, Lower Cambrian, time equivalent to *Protolenus* Zone (Vidal & Peel 1993).

Heliosphaeridium obscurum (Volkova, 1969) Moczydłowska, 1991

Synonymy and description. – See Moczydłowska 1991.
Additionally: □*non* 1986 *Micrhystridium obscurum* Volkova 1969 – Welsch, pp. 55–56, Pl. 3:1–6. □1989 *Micrhystridium obscurum* Volkova 1969b – Hagenfeldt 1989a,
pp. 80–82, Pl. 4:2. □1994 *Micrhystridium obscurum* Volkova 1969 – Sarjeant & Stancliffe, p. 17.

Material. – Four poorly preserved specimens.

Dimensions. – *N*=3. Vesicle diameter 10–20 μm, length of processes 2–5 μm.

Present record. – The Upper Silesia area, Goczałkowice IG-1 borehole, Goczałkowice formation at a depth of 2854.4–2862.0 m, Lower Cambrian, *Holmia kjerulfi* Zone; Sosnowiec IG-1 borehole, Sosnowiec formation at a depth of 3403.5–3407.2 m, Middle Cambrian, *Acadoparadoxides oelandicus* Superzone.

Occurrence and stratigraphic range. – The species occurs in Poland, Latvia, Lithuania, Estonia, Sweden, Finland, Belgium, Canada, Turkey and Kazakhstan in Lower Cambrian, equivalent to *Holmia kjerulfi* and *Protolenus* Zones, and Middle Cambrian, *Acadoparadoxides oelandicus* Superzone (see comprehensive summaries by Hagenfeldt 1989a and Moczydłowska 1991). The species has also been reported from Spain, Cantabrian Mountains at Barrios de Luna, Herreria Formation, Lower Cambrian, equivalent to *Holmia kjerulfi* and *Protolenus* Zones (Palacios & Vidal 1992). Greenland, Peary Land, Buen Formation, Lower Cambrian, equivalent to *Holmia kjerulfi* and *Protolenus* Zones (Vidal & Peel 1993).

Heliosphaeridium oligum (Jankauskas, 1976) n.comb.

Fig. 31I–K

Synonymy. – □1976 *Micrhystridium oligum* Jankauskas, sp. nov – Jankauskas & Posti, p. 147, Pl., Fig. 10, 13, 15, 16, 20. □1979 *Micrhystridium oligum* Jankauskas, 1976 – Volkova *et al.*, p. 15, Pl. 9:8–10. □non 1982 *Micrhystridium* cf. *oligum* Jankauskas – Downie, Fig. 6ee.

Material. – Six well-preserved specimens.

Description. – Vesicles circular in outline, possessing scarce (15 observed on the vesicle contour), evenly distributed processes. The processes are short, of equal length and slightly widened at the base. The terminations of the processes are blunt or sligtly widened, occasionally funnel-like or divided. The processes are hollow (at least in their basal portions), and their cavities communicate with the inner cavity of the vesicle.

Dimensions. – *N*=6. Diameter of vesicle 10–15 μm, length of processes 2–3 μm.

Remarks. – In the original diagnosis of the species *Micrhystridium oligum* (Jankauskas in Jankauskas & Posti 1976), the nature of the processes was not defined. Because of their small dimensions, it is difficult to observe whether or not the processes are hollow in their full length. It seems that the bases of the processes have inner cavities.

Present record. – The Upper Silesia area, Sosnowiec IG-1 borehole, Sosnowiec formation at a depth of 3365.2–3372.3 m, Middle Cambrian, *Acadoparadoxides oelandicus* Superzone.

Occurrence and stratigraphic range. – Lithuania, borehole Kibartai-22, Lower Cambrian, Rausve horizon (Jankauskas & Posti 1976; Volkova *et al.* 1979).

Heliosphaeridium serridentatum n.sp.

Fig. 32I–K

Holotype. – Specimen PMU-Pl.90-X/38/3; Fig. 32I.

Synonymy. – 1986 *Micrhystridium brevicornum* Jankauskas, 1976 – Welsch, pp. 52–53, Pl. 2:16–20.

Derivation of name. – From Latin *serra* – saw, and *dens* – tooth, in reference to the saw-toothed shape of the processes.

Locus typicus. – The Upper Silesia area, Sosnowiec IG-1 borehole (Figs. 2 and 3).

Stratum typicum. – Alternating mudstones and fine-grained sandstones in the Sosnowiec formation at a depth of 3365.2–3372.3 m; Middle Cambrian, *Acadoparadoxides oelandicus* Superzone.

Material. – Seven satisfactorily preserved specimens.

Diagnosis. – Organic-walled microfossils with vesicles that are circular to oval in outline, originally spherical, bearing numerous (approximately 25–30 observed on vesicle outline) short and rigid processes that are evenly distributed on the vesicle. The wall of the vesicle is thick. The processes are simple, thorn-like with wide bases forming a saw-toothed outline of the vesicle. The processes are hollow and connected with the inner cavity of the vesicle, but their distal portions may be solid. The tips of the processes are sharp-pointed, occasionally blunt.

Dimensions. – *N*=7. Vesicle diameter 16–23 μm (holotype 18 μm), length of processes around 2 μm.

Remarks. – Acritarchs from Finnmark, Norway, attributed to *Micrhystridium brevicornum* Jankauskas, 1976 (Welsch 1986, pp. 52–53, Pl. 2:16–20) are considered conspecific with *Heliosphaeridium serridentatum* n.sp. The specimens from Finnmark differ considerably from *M. brevicornum*, as described and illustrated by Jankauskas (1976, p. 189, Pl. 25:9, 12; figured specimens were erroneously numbered in the original publication, cf. Volkova *et al.* 1979, p. 13, Pl. 9:21–24). *M. brevicornum* has rare and irregularly distributed processes, in groups or rows (Jankauskas 1976, p. 189). However, specimens of *M. brevicornum* illustrated by Jankauskas (1976, Pl. 25:9, 12) and Volkova *et al.* (1979, Pl. 9:21–24) are poorly preserved.

Their processes seem to be heteromorphic and randomly distributed; this rather looks like the result of the state of preservation. Most likely, taphonomic features were incorporated in the diagnosis of the species. These circumstances render recognition of *M. brevicornum* as a distinct species difficult.

Microfossils recovered from the Kistedal Formantion in Finnmark have numerous, evenly distributed, short and robust processes that are closely arranged. The processes have wide bases and blunt tips, and the vesicle wall is thick and rigid (Welsch 1986). The illustrated specimens have processes with both blunt and sharp tips, but a significant feature is that they are regular, rigid and thornlike. The specimens are strongly thermally altered and as result carbonized and thus opaque. This state of preservation rendered it impossible to observe whether the processes are hollow or solid (Welsch 1986). The overall morphologic features, however, correspond to the morphology of the new species from Silesia, and they are regarded as conspecific.

Present record. – As for the holotype.

Occurrence and stratigraphic range. – Norway, Finnmark, Digermul Peninsula, lower part of the Kistedal Formation, Middle Cambrian, *Paradoxides paradoxissimus* Zone (Welsch 1986).

Heliosphaeridium sp.

Material. – Eight poorly preserved specimens.

Present record. – The Upper Silesia area, Goczałkowice IG-1 borehole, Goczałkowice formation at a depth of 2771.8–2789.0 m, Lower Cambrian, probably *Protolenus* Zone; Sosnowiec IG-1 borehole, Sosnowiec formation at depths of 3365.2–3372.3 m and 3210.0–3212.0 m, Middle Cambrian, *Acadoparadoxides oelandicus* and *Paradoxides paradoxissimus* Superzones, respectively.

Genus *Leiosphaeridia* Eisenack, 1958, emended Downie & Sarjeant, 1963

Type species. – *Leiosphaeridia baltica* Eisenack, 1958a, p. 8, Pl. 2:5; Estonia, Ordovician, Ashgill.

Leiosphaeridia sp.

Figs. 20I, 27I

Material. – Very abundant specimens in variable states of preservation.

Remarks. – Individual species are not recognized among the Cambrian spheromorphic microfossils attributed to the genus *Leiosphaeridia*, because of the lack of objective diagnostic characters (Moczydłowska 1991, p. 61). These microfossils have no stratigraphic significance. Another spheromorphic genus, *Dichotisphaera* Turner, 1984, is considered a junior synonym of *Leiosphaeridia* (Le Hérissé 1989, p. 63). According to Turner (1984, pp. 107–108) *Dichotisphaera* differs from *Leiosphaeridia* by having a median split as excystment structure, being a diagnostic character, contrary to the pylome in the latter genus (Eisenack 1958a; Downie & Sarjeant 1963). The excystment opening alone is a poor diagnostic feature, because most microfossils do not display any, being preserved in non-excysting stages of the life cycle of the microorganisms that produced them. Some of the ruptures on the wall of vesicles are also taphonomic (preservational) and are impossible to distinguish from the true excystment openings by splitting.

Accordingly, *Dichotisphaera gregalis* (Hagenfeldt, 1989a) Vanguestaine, 1991, recently recognized in the Cambrian (Vanguestaine 1991, Martin *in* Young *et al.* 1994) is here considered to be undifferentiated species of *Leiosphaeridia*. This species was erected as *Leiosphaeridia gregalis* (Hagenfeldt 1989a, Pl. 3:5, 6) though neither the diagnosis nor the photomicrograph of the holotype provide any diagnostic features that would allow the 'species' to be distinguished from the other simple spheromorphic acritarchs. *Leiosphaeridia* sp. 1 by Moczydłowska & Vidal (1986) was regarded as a senior synonym of *L. gregalis*, and its stratigraphic range and distribution were included in those of *L. gregalis* (Hagenfeldt 1989a, p. 62). *Leiosphaeridia* sp. 1 lacks morphological traits that would enable it to be distinguished as a separate species, and it has no particular stratigraphic significance, since such spheromorphs occur throughout the Neoproterozoic–Cambrian (Moczydłowska 1991, p. 61). Vanguestaine (1991, p. 218) subsequently transferred *L. gregalis* to the genus *Dichotisphaera* Turner, 1984. However, the Ordovician microfossil selected as the type species for *Dichotisphaera*, *D. caradociensis* (Turner 1984, p. 108, Pl. 15:7, 8), is indistinguishable by any morphological means from specimens referred to *L. bicrura* Jankauskas, 1976 (*in* Jankauskas & Posti 1976, Pl., Fig. 11, 21), *L. dehisca* Paškevičiene, 1979 (*in* Volkova *et al.* 1979, Pl. 31:7, 8, 11, 17–19, 22) or *L. gregalis* Hagenfeldt, 1989a (1989a, Pl. 3:5, 6). All these species represent different states of preservation of spheromorphic microfossils displaying ruptures on the vesicle wall. These ruptures, as observed, were not morphologically predetermined, and so, whether they are in the central part of vesicle (then called median split) or asymmetrically located (then partial rupture or partial splitting), they are random features. None of the above species is morphologically and objectively recognizable or restricted to particular stratigraphic intervals. Therefore,

Locus typicus. – The Upper Silesia area, the Sosnowiec IG-1 borehole (Figs. 2 and 3).

Stratum typicum. – A mudstone interbed in fine-grained sandstones of the Sosnowiec formation at a depth of 3166.3–3167.7 m, Upper Cambrian.

Material. – Seventeen specimens in various states of preservation.

Diagnosis. – Vesicle originally spherical, circular to oval in outline, single-layered, bearing long cylindrical processes that branch distally. The surface of the wall and processes is smooth to microgranulate. The processes are evenly distributed and not very numerous (approximately 12–16 observed on the vesicle outline), slightly conical at the bases or arising perpendicularly from the vesicle wall. The dichotomizing terminations of processes form a small corona of second and third order pinnulae. The processes are hollow and connected freely with the vesicle cavity.

Dimensions. – $N=10$. Vesicle diameter 27–43 μm (holotype 40×49 μm), length of processes 9–18 μm (holotype 14–18 μm).

Remarks. – The basal parts of some processes are darker and more dense in appearance and may appear to have a plug. This is caused by the compaction and condensation processes in organic matter during fossilization and depends on the state of preservation of microfossils.

The new species is distinct from other species of *Multiplicisphaeridium* by having a relatively small corona of pinnulae and long tubular processes.

Present record. – As for the holotype.

Multiplicisphaeridium varietatis n.sp.

Fig. 37A–D

Holotype. – Specimen PMU-Pl.115-W/44/1; Fig. 37A. Paratype specimen PMU-Pl.116-W/42/2; Fig. 37B.

Derivation of name. – From Latin *varietas* – variety, diversity, referring to the various shapes of the branching processes and the process bases.

Locus typicus. – The Upper Silesia area, the Sosnowiec IG-1 borehole (Figs. 2 and 3).

Stratum typicum. – A mudstone interbed in fine-grained sandstones of the Sosnowiec formation at a depth of 3166.3–3167.7 m, Upper Cambrian.

Material. – Twenty-eight fairly well-preserved specimens.

Diagnosis. – Vesicle originally spherical, circular to oval in outline, bearing not very numerous (approximately 15–18 observed on the vesicle outline), cylindrical processes with ramifying terminations. The surface of the vesicle

wall, and occasionally that of the processes, is granular. The processes have conical bases which are transitionally elongated into cylindrical trunks or which may be contracted and plugged. Both kinds of basal portions of processes may occur on the same specimen. The contracted bases, if present, are narrow and opaque. The processes are hollow, and their cavities communicate with the vesicle, or are separated by plug. The distal terminations of processes are variably ramified, up to the third order of pinnulae, forming large coronae. The tips of the pinnulae are closed and acuminate. The opening (pylome) is large and irregular, having a denticular outline formed by the granular sculpture.

Dimensions. – $N=8$. Vesicle diameter 35–50 μm (holotype 40×42 m), length of processes 9–21 μm (holotype 13–21 μm), width of processes around 3 μm, spread of corona of pinnulae is 8–12 μm.

Remarks. – The shape of the processes in the new species sometimes resembles that of *Timofeevia lancarae*, but the major difference is, at the generic level, the lack of polygonal fields on the vesicle and crests between the processes that are characteristic of *Timofeevia*.

Occasionally, a few wrinkles may occur on the vesicle wall around the base of processes. Sometimes three differently preserved process bases occur on a single specimen: freely communicating with the vesicle, with secondary plug, and with the radial wrinkles surrounding the base. These taphonomic features are clearly related to the state of preservation of the microfossils. Well-preserved specimens display convex and undivided vesicle wall and the process cavities open into the interior of the vesicle.

Present record. – As for the holotype, and Potrójna IG-1 borehole, Jaszczurowa formation at a depth of 3356.3–3363.5 m, Upper Cambrian.

Multiplicisphaeridium xianum Fombella, 1977

Fig. 34D–F

Synonymy. – □*non* 1976 *Multiplicisphaeridium dendroideum* (Burmann 1970) Eisenack *et al.*, pp. 455–456. □1976 *Baltisphaeridium dendroideum* Jankauskas, sp.nov. – Jankauskas, p. 189, Pl. 25:20. □1977 *Multiplicisphaeridium xianum* Fombella n.sp. – Fombella, p. 119, Pl. 1:13, Text-fig. 1:10. □invalid senior homonym 1979 *Multiplicisphaeridium dendroideum* (Jankauskas, 1976) Jankauskas et Kirjanov, comb.nov. – *in* Volkova *et al.*, pp. 17–18, Pl. 3:1–7. □1982 *Multiplicisphaeridium dendroideum* (Jankauskas 1976) Volkova *et al.* – Downie, p. 262, Fig. 7k–l. □1989 *Multiplicisphaeridium dendroideum* (Yankauskas) Yankauskas and Kirjanov, 1979 – Hagenfeldt, 1989a, pp. 97–99, Pl. 4:5. □1990 *Multiplicisphaerid-*

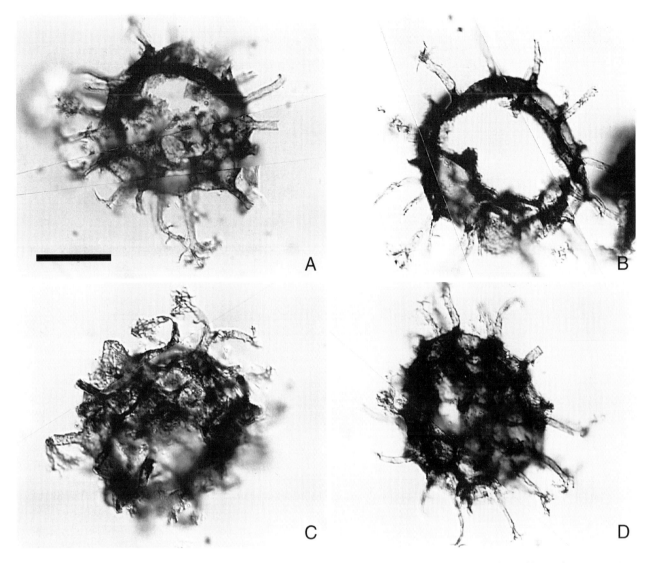

Fig. 37. □A–D. *Multiplicisphaeridium varietatis* n.sp.; A, PMU-Pl.115-W/44/1, holotype, hollow tubular processes with variable ramifying terminations forming a corona of pinnulae; note granular sculpture on the vesicle wall and processes; B, PMU-Pl.116-W/42/2, paratype, with large pylome having denticular sculpture around the rim; the process bases are contracted and plugged, a taphonomic feature; C, PMU-Pl.117-Y/31/1, free connection between the process and vesicle cavities is exhibited on the left side of the specimen, and bifurcated trunk of process at the top; D, PMU-Pl.118-T/42/1, showing plugs at process bases. Specimens in A, B, D from the Sosnowiec IG-1 borehole, depth 3166.3–3167.7 m, Sosnowiec formation, lower Upper Cambrian; C from the Potrójna IG-1 borehole, depth 3356.3–3363.5 m, Jaszczurowa formation, Upper Cambrian. Scale bar in A equals 20 μm for all micrographs.

ium yankauskasii nom.nov. *subst. pro Multiplicisphaeridium dendroideum* (Yankauskas 1976) Yankauskas and Kiryanov *in* Volkova *et al.* 1979 – Fensome *et al.*, p. 358. □1991 *Multiplicisphaeridium dendroideum* (Jankauskas, 1976) Jankauskas & Kirjanov, 1979 – Moczydłowska, p. 63, Pl. 9L. □1993 *Multiplicisphaeridium dendroideum* (Jankauskas, 1976) Jankauskas & Kirjanov, 1979 – Vidal & Peel, p. 29, Fig. 11b, c. □1994 *Multiplicisphaeridium dendroideum* (Jankauskas) Jankauskas & Kirjanov *in* Volkova *et al.* 1979 – Martin *in* Young *et al.* Fig. 10a.

Material. – Twelve well-preserved specimens.

Description. – Vesicles circular to oval in outline, originally spherical, bearing long and robust processes. The processes are wide proximally and gradually tapering distally. The distal ends of the processes are heteromorphic, simple and divided into two or three branches. The cavity of the central body is connected with the inner cavity of the processes.

Dimensions. – $N = 10$. Diameter of central body 9–19 μm, length of processes 4–9 μm.

Remarks. – *Multiplicisphaeridium xianum* Fombella, 1977, is a senior, validly published synonym of *Multiplici-*

Polygonium sp. A

Fig. 45C

Material. – Six fairly well-preserved specimens.

Description. – Vesicle ovoid in outline, having a central opaque body and transparent, relatively short processes. The processes are conical with conspicuous, wide and bulbous bases, and taper towards sharp-pointed terminations. The cavities of the central body and processes are freely connected. The bases of processes are closely arranged, forming an undulating outline of the vesicle. Eight to ten processes are observed on the vesicle contour.

Dimensions. – $N=2$. Vesicle diameter 14×18 μm and 14×23 μm, length of processes 5–8 μm.

Remarks. – The described specimens display morphological features, such as bulbous process bases and variable proportions of processes, that distinguish them from the previously recognized species of *Polygonium*, and they probably belong to a new species.

Present record. – The Upper Silesia area, Goczałkowice IG-1 borehole, Goczałkowice formation at a depth of 2766.8–2771.1 m, Middle Cambrian, *Acadoparadoxides oelandicus* Superzone; Sosnowiec IG-1 borehole, Sosnowiec formation at a depth of 3365.2–3372.3 m, Middle Cambrian, the *Acadoparadoxides oelandicus* Superzone, and at a depth of 3174.0–3174.7 m, lower Upper Cambrian.

Genus *Pterospermella* Eisenack, 1972

Type species. – *Pterospermella aureolata* (Cookson & Eisenack) Eisenack, 1972, p. 597, Text-figs. 1–3 (=*Pterospermopsis aureolata* Cookson & Eisenack, 1958, p. 49, Pl. 9:10–12); Australia, Cretaceous.

Pterospermella vitalis Jankauskas, 1979

Fig. 27H

Material. – Three poorly preserved specimens.

Synonymy and description. – See Jankauskas in Volkova *et al.* (1979) and Moczydłowska 1991, p. 64.

Dimensions. – $N=3$. Overall diameter of vesicle 18–25 μm, diameter of inner body 13–16 μm, width of membrane 4–5 μm.

Present record. – The Upper Silesia area, Sosnowiec IG-1 borehole, Sosnowiec formation at a depth of 3365.2–3372.3 m; Middle Cambrian, *Acadoparadoxides oelandicus* Superzone.

Occurrence and stratigraphic range. – Lithuania, Lower Cambrian, Vergale horizon and Middle Cambrian, Kibartai horizon; Estonia, Lower Cambrian, Talsy horizon (Volkova *et al.* 1979). Poland, the East European Platform, Mazowsze Formation, Lower Cambrian, *Platysolenites* Zone, Zawiszyn Formation, Lower Cambrian, *Schmidtiellus mickwitzi* Zone, Kaplonosy and Radzyń formations, Lower Cambrian, *Holmia kjerulfi* Zone (Volkova *et al.* 1979; Moczydłowska 1991).

Pterospermella sp.

Material. – Five poorly preserved specimens.

Present record. – The Upper Silesia area, Sosnowiec IG-1 borehole, Sosnowiec formation at a depth of 3210.0–3212.0 m, Middle Cambrian, *Paradoxides paradoxissimus* Superzone, and at a depth of 3197.5–3198.5 m, Middle Cambrian, probably *Paradoxides forchhammeri* Superzone.

Genus *Retisphaeridium* Staplin *et al.*, 1965

Type species. – *Retisphaeridium dichamerum* Staplin *et al.*, 1965, pp. 187–188, Pl. 19:3 (holotype) and Pl. 19:1–7; Canada, Alberta at the California Standard Parkland, Middle Cambrian.

Retisphaeridium brayense (Gardiner & Vanguestaine, 1971) Moczydłowska & Crimes, 1995

Synonymy and description. – See Moczydłowska & Crimes 1995, pp. 122–123.

Material. – Twelve poorly preserved specimens.

Dimensions. – $N=8$. Vesicle diameter 25–37 μm.

Present record. – The Upper Silesia area, Sosnowiec IG-1 borehole, Sosnowiec formation at a depth of 3203.5–3205.5 m, Middle Cambrian, probably *Paradoxides forchhammeri* Superzone.

Occurrence and stratigraphic range. – The species has been found in the Middle and Upper Cambrian strata in Ireland, Belgium, France, Italy, Czech Republic, Canada (Newfoundland), and Tunisia (see compilation and the age re-assessment for the record in Ireland *in* Moczydłowska & Crimes 1995).

Genus *Revinotesta* Vanguestaine, 1974, emended

Type species. – *Revinotesta microspinosa* Vanguestaine, 1974, p. 74, Pl. 1:8; Belgium, Stavelot Massif, Middle Cambrian (Vanguestaine 1974, 1992).

Synonymy. – □invalid name 1975 *Ovulum* Jankauskas, gen.nov. – Jankauskas, p. 96. □1975 *Aranidium* Jankauskas, gen.nov. – Jankauskas, p. 99.

Emended diagnosis. – Organic-walled microfossils with minute vesicles that are ovoid, ellipsoidal to circular in outline, having a circular opening (aperture) at the apical end of the vesicle. The wall of the vesicle is smooth or ornamented with granulae, slightly elongated solid spines, or clavate ornaments. The opening of the vesicle may have an even or irregular edge or may be surrounded by ornamental elements. The operculum, occasionally present, is formed by part of the vesicle wall detached along the edge of the aperture.

Remarks. – The genus *Revinotesta* Vanguestaine, 1974 (Vanguestaine 1974, p. 73), includes minute microfossils with an apical aperture whose vesicle wall may have no ornamentation (so-called 'reduced ornamentation') or may possess spines on the wall and around the opening. This concept of the genus allows incorporation of *Aranidium* Jankauskas, 1975, as a junior synonym, and *Ovulum* Jankauskas, 1975, as an invalid junior synonym (Cramer & Diez 1979, pp. 43, 94; Fensome *et al.* 1990, pp. 62, 378, 440). In the original diagnosis of *Aranidium*, the existence of a vesicle opening was not mentioned, reference being made only to small ovoid, flattened vesicles with spines (Jankauskas 1975). However, the type species (*A. izhoricum* Jankauskas, 1975) and one additional species (*A. aculeatum* Jankauskas, 1975, here regarded as its synonym) obviously possess a wide opening (called a pylome by Jankauskas 1975, p. 99) at one of the poles of the vesicles. Later, the diagnosis of *Aranidium* was extended by adding reference to the presence of an opening (then called 'pore?') surrounded by a ring of spines forming a 'neck' (Volkova *et al.* 1979). This 'neck' was indicated as a diagnostic feature of *Aranidium* to differentiate it from the genus *Revinotesta* Vanguestaine, 1974 (Volkova *et al.* 1979, p. 30). This is, however, not an obvious dissimilarity, and the presence of a 'neck' was convincingly demonstrated only in *Aranidium sparsum* Volkova, 1979 (in Volkova *et al.* 1979), a species that was erected subsequent to the genus. This feature is rather insignificant and may serve to distinguish species within a genus, instead of differentiating between separate genera.

　Most of the species assigned to *Aranidium* by Jankauskas (1975) were described as having no opening (e.g., *A. undosum*, *A. obsoletum*, *A. confusum* and *A. pycnacanthum*). Subsequently, it was established that *A. pycnacan-*

thum possesses a pylome, whereas three former species could be excluded from the genus (Volkova *et al.* 1979, p. 30). Morever, the apical opening surrounded by spines was believed not to be an excystment opening (and therefore not a pylome) but the exit opening for one or several undulipodia (Volkova *et al.* 1979). This point of view has been contradicted by Welsch (1986, pp. 88–89, Pl. 1:1–5) who demonstrated that microfossils having an opening with distinctive granular marginal ornamentation may also possess an operculum *in situ*, thus representing an excystment structure. He considered that the opening undoubtedly served to excyst the vesicle content and that it opened only once. This type of opening differs only in morphology from the randomly located circular pylome (genus *Peteinosphaeridium*), irregularly shaped pylome (*Stelliferidium*, *Multiplicisphaeridium*), or irregular rupture (*Skiagia*, *Gorgonisphaeridium*, *Baltisphaeridium*), but all of them functioned as the excystment structure of the resting cysts of planktic algal protists (Vavrdová 1976; Tappan 1980; Mendelson 1993).

　Some of the species attributed to *Aranidium* and *Ovulum* are considered to be synonymous and are herein transferred to *Revinotesta*. A revised list of species of the genus *Revinotesta* is proposed as follows:

Revinotesta microspinosa Vanguestaine, 1974, by original diagnosis.

Revinotesta izhorica (Jankauskas, 1975) n.comb., having opening surrounded by short, stiff spines and short spines densely distributed over the vesicle. It includes the synonymous species *Aranidium aculeatum* Jankauskas, 1975, and *A. pycnacanthum* Jankauskas, 1975.

Revinotesta ordensis Downie, 1982, emend., having long and slender spines.

Revinotesta sparsa (Volkova, 1979) n.comb., having long spines on the vesicle with slightly widened conical bases and short spines bordering the opening (*in* Volkova *et al.*, 1979).

Revinotesta saccata (Jankauskas, 1975) *ex* Fensome *et al.*, 1990, having smooth or granular ovoidal vesicle with simple circular or irregularly shaped opening.

Revinotesta lanceolata (Jankauskas, 1975) *ex* Fensome *et al.*, 1990, having ovoidal or ellipsoidal, smooth-walled vesicle with a simple opening on one end of the vesicle and an elongated lanceolate-shaped opposite end with a short single spine.

Revinotesta comosa (Vavrdová, 1984) n.comb., (=*Aranidium comosum*, Vavrdová 1984, pp. 168–169, Pls. 1:2, 3; 4:4–6, Text-fig. 4a, b), by original diagnosis.

Aranidium granulatum Welsch, 1986, should be transferred to *Heliosphaeridium*, as *Heliosphaeridium granulatum* (Welsch, 1986) n.comb. The species has regular hollow processes rather than solid ornamentation spines (Welsch 1986, Pl. 1:1–5). The presence of a circular opening surrounded by coarse granulae and having an operculum is not restricted to the genus *Aranidium* (here = *Revinotesta*).

Species of *Aranidium* without an opening, described by Jankauskas (1975), should be transferred to other genera. *Aranidium confusum* (Jankauskas 1975, pp. 102–103, Pl. 11:59, 60, 62–65), referred to *Micrhystridium* by Fensome *et al.* (1990, p. 63), is here regarded as belonging to *Heliosphaeridium* Moczydłowska, 1991. *Aranidium undosum* (Jankauskas 1975, p. 101, Pl. 11:54) is probably conspecific with *A. confusum*. *Aranidium obsoletum* (Jankauskas 1975, p. 102, Pl. 11:44–46) most likely belongs to the genus *Heliosphaeridium*.

The species *Revinotesta punctata* Tynni, 1982, *ex* Fensome *et al.*, 1990 (= *Ovulum punctatum* n.sp., Tynni 1982, p. 53, Fig. 13T), listed among species of *Revinotesta* by Fensome *et al.* (1990, pp. 378 and 440), is not considered here as a formal species. The reason is that 'Ovulum punctatum n.sp.' was not validly published because the holotype of the species was not indicated. The photomicrograph of 'Ovulum punctatum' (Tynni 1982, Fig. 13T) was not indicated as the holotype, nor was information about the provenance of the specimen provided. Besides the non-compliance with these formal requirements, the illustrated specimen displays no diagnostic features other than those attributable to the alteration of organic matter. Fensome *et al.* (1990, p. 440) regarded 'Ovulum punctatum' as invalidly published because the generic name was not validly published, and they placed this microfossil in the genus *Revinotesta*. They also regarded the species to be Ordovician in age. This could not be properly established, as the sample consisted of mixed assemblages of microfossils of early Cambrian and early Ordovician ages (Tynni 1982, p. 56).

Revinotesta izhorica (Jankauskas, 1975) n.comb.

Fig. 31N

Synonymy. – □1975 *Aranidium izhoricum* Jankauskas, sp.nov. – Jankauskas, p. 99, Pl. 11:25–32. □1975 *Aranidium aculeatum* Jankauskas, sp.nov. – Jankauskas, pp. 99–101, Pl. 11:33–39. □1975 *Aranidium pycnacanthum* Jankauskas, sp.nov. – Jankauskas, pp. 101–102, Pl. 11:40–43. □1982 *Aranidium* cf. *izhoricum* Jankauskas – Downie 1982, Fig. 10h.

Material. – Twelve poorly preserved specimens.

Description. – Minute vesicles, ovoid in outline, having a circular opening at the apical part of the vesicle and bearing numerous spines evenly distributed on the vesicle and around the opening. The spines are stiff in appearance,

thorn-like and solid. The basal portions of the spines are slightly widened and the tips are sharp-pointed.

Dimensions. – N = 5. Diameter of vesicle 7–9 μm, diameter of opening 4–5 μm, length of spines 1–2 μm.

Remarks. – Two species of *Aranidium* recognized by Jankauskas (1975), *A. aculeatum* and *A. pycnacanthum*, are here considered to be conspecific with the type species *A. izhoricum* = *Revinotesta izhorica* (Jankauskas, 1975) n.comb. All these species have similar morphology and dimensions (cf. Jankauskas 1975; Volkova *et al.* 1979) which do not allow their objective specific distinction.

Present record. – The Upper Silesia area, Sosnowiec IG-1 borehole, Sosnowiec formation at depths of 3403.5–3407.2 m and 3365.2–3372.3 m, Middle Cambrian, *Acadoparadoxides oelandicus* Superzone; at a depth of 3174.0–3174.7 m, lower Upper Cambrian.

Occurrence and stratigraphic range. – Lithuania, Latvia and Russia, Lower Cambrian, Vergale horizon (Jankauskas 1975). Scotland, Skiag Bridge, 'Fucoid Beds', Lower Cambrian, time equivalent to *Holmia kjerulfi* Zone (Downie 1982).

Revinotesta microspinosa Vanguestaine, 1974

Fig. 31M

Synonymy. – □1974 *Revinotesta microspinosa* n.sp. – Vanguestaine, p. 74, Pl. 1:8. □1978 *Revinotesta microspinosa* Vanguestaine, 1974 – Vanguestaine, Pl. 2:18–26. □1980 *Aranidium* sp. – Volkova, Pl., Fig. 16. □1981 *Revinotesta microspinosa* Vanguestaine, 1974 – Erkmen & Bozdoğan, p. 54, Pl. 1:24. □1990 *Aranidium* sp. – Volkova, Pls. 2:10; 5:14.

Material. – Four poorly preserved specimens.

Description. – Minute vesicles, ellipsoidal in outline, having a circular opening on the apical part of the vesicle and bearing a few short and solid spines. The spines are distributed on the wall of the vesicle and in proximity to the opening.

Dimensions. – N = 4. Diameter of vesicle 5–9 μm, length of spines less than 1 μm.

Present record. – The Upper Silesia area, Sosnowiec IG-1 borehole, Sosnowiec formation at a depth of 3403.5–3407.2 m; Middle Cambrian, *Acadoparadoxides oelandicus* Superzone.

Occurrence and stratigraphic range. – Belgium, Stavelot Massif and Givonne Massif, and France, the Ardennes, Middle Cambrian (Vanguestaine 1974, 1978, 1992). SE Turkey, Mardin–Derik area, Sosink Formation, Middle

Cambrian (Erkmen & Bozdoğan 1981), equivalent to *Paradoxides paradoxissimus* Superzone. Russia, Moscow syneclise, Tolbukhino borehole, Sablinsk Formation, Middle Cambrian (Volkova 1980); Jaroslav district, Danilovsk-11 borehole, Molozhsk Formation, upper Middle Cambrian – lower Upper Cambrian (Volkova 1990).

Revinotesta ordensis Downie, 1982, emended

Fig. 31O–P

Synonymy. – □*nomen nudum* 1979 *Aranidium delicatum* n.sp. – Fombella, p. 4, Pl. 2:31. □1979 *Aranidium* sp. – Volkova *et al.*, Pl. 11:8–12, 20. □1982 *Revinotesta ordensis* sp.nov. – Downie, p. 265, Fig. 10g, i–k.

Material. – Seven well-preserved specimens.

Emended diagnosis. – Minute vesicles, ellipsoidal in outline, having a circular opening on the apical part of the vesicle and bearing rare simple spines on the surface of the vesicle. The spines are solid, very thin and fragile, appear to be stiffer proximally, and have sharp tips.

Dimensions. – $N=3$. Diameter of vesicle 6–11 μm, length of spines 3–4 μm.

Remarks. – *Revinotesta ordensis* was diagnosed as having blunt or capitate tips of the spines (Downie 1982, p. 265). This is, however, not demonstrated in light photomicrographs illustrating the holotype, nor in additional specimens of the species (Downie 1982, Fig. 10i, j, k, l), whose spines are very thin and clearly tapering towards their terminations. Blunt or capitate process tips are shown on the SEM photomicrograph of a microfossil referred to as *R. ordensis* (Downie 1982, Fig. 10g), but its processes are cylindrical and it might belong to a different species. Therefore, the diagnosis of *R. ordensis* is here emended in accordance with features shown on the holotype of the species, enabling the practical recognition of the species under the light microscope. The present specimens, displaying sharp-pointed tips under the light microscope and having similar overall appearance to the holotype, are attributed to *R. ordensis* Downie, 1982, emended.

The species differs from *Revinotesta microspinosa* Vanguestaine, 1974, by having much longer and more flexible spines.

The species *Aranidium elicatum* is a *nomen nudum*, since a holotype was never described or validly published (Fombella 1979, p. 4; Fensome *et al.* 1990, p. 63).

Present record. – The Upper Silesia area, Sosnowiec IG-1 borehole, Sosnowiec formation at depths of 3403.5–3407.2 m, 3365.2–3372.3 m and 3174.0–3174.7 m; Middle Cambrian, *Acadoparadoxides oelandicus* Superzone and lower Upper Cambrian, respectively.

Occurrence and stratigraphic range. – Spain, the Cantabrian Mountains at Lois, the Oville Formation, lower Middle Cambrian (Fombella 1979). Scotland, Skiag Bridge, Fucoid Beds, Lower Cambrian, equivalent to the *Holmia kjerulfi* Zone (Downie 1982). Latvia, Lower Cambrian, Rausve horizon, and Middle Cambrian, Kibartai horizon (Volkova *et al.* 1979).

Revinotesta saccata (Jankauskas, 1975) n.comb.

Synonymy. – □invalid 1975 *Ovulum saccatum* Jankauskas, sp.nov. – Jankauskas, pp. 96–97, Pl. 11:1–15, 23. □invalid 1979 *Ovulum saccatum* – Volkova *et al.*, p. 31, Pl. 11:22–27. □1982 *Ovulum saccatum* Jankauskas 1975 – Downie, p. 265, Fig. 10m. □1990 *Ovulum* sp. – Volkova, Pls. 1:11; 5:8–9, 11–13.

Material. – Four poorly preserved specimens.

Description. – Minute vesicles, ovoid in outline, having a circular opening at the apical pole of the vesicle. The wall of the vesicle is smooth, occasionally ornamented by rare granular elements.

Dimensions. – $N=4$. Diameter of vesicle 16–23 μm, diameter of opening 5–6 μm.

Remarks. – Granulae on the vesicle wall are rare and irregularly distributed, resulting in some portions of the vesicle surface being smooth. It is difficult to assess if this is a morphological pattern of random distribution or the result of taphonomy. Specimens with simple morphology and inconspicuous granular ornamentation were illustrated by Volkova (1990) as *Ovulum* sp.

Present record. – The Upper Silesia area, Sosnowiec IG-1 borehole, Sosnowiec formation at a depth of 3403.5–3407.2 m; Middle Cambrian, *Acadoparadoxides oelandicus* Superzone.

Occurrence and stratigraphic range. – Lithuania, Latvia, Estonia and the Ukraine, Lower Cambrian, Vergale and Rausve horizons; Latvia, Middle Cambrian, Kibartai horizon (Jankauskas 1975; Volkova *et al.* 1979). Russia, Jaroslav area, Danilov-11 and Tolbukhino-1 boreholes, Molozhsk Group, and St. Petersburg area, Zareche borehole, Petseriy beds, upper Middle Cambrian – lower Upper Cambrian; Kaliningrad area, Veselovsk-5 borehole, the Veselovsk Formation, Middle Cambrian; St. Petersburg area, Izhora River, the Ladoga Formation, Upper Cambrian, *Leptoplastus* and *Peltura* Zones (Volkova 1990). Scotland, Skiag Bridge, 'Fucoid Beds', Lower Cambrian, time-equivalent to *Holmia kjerulfi* Zone (Downie 1982).

Genus *Skiagia* Downie, 1982, emended Moczydłowska, 1991

Type species. – *Skiagia scottica* Downie, 1982, p. 264, Figs. 5, 8k–l, 9a–f, holotype 9c; Scotland, Fucoid Beds, Lower Cambrian, time equivalent to the *Holmia kjerulfi* Zone.

Skiagia ciliosa (Volkova, 1969) Downie, 1982

Fig. 39A–B

Synonymy and description. – See Moczydłowska 1991, pp. 65–66. Additional: □*non* 1989 *Skiagia ciliosa* (Volkova) Downie, 1982, forma 1 – Hagenfeldt 1989a, Pl. 4:12. □1989 *Skiagia ciliosa* (Volkova) Downie, 1982, forma 2 – Hagenfeldt 1989a, pp. 108–109, Pl. 5:1. □1989 *Skiagia compressa* (Volkova) Downie, 1982 – Hagenfeldt 1989a, Pl. 5:2. □1989 *Skiagia ciliosa* (Volkova) Downie, 1982 forma 3 – Hagenfeldt 1989b, pp. 224–225, Pl. 3:1. □1992 *Skiagia ciliosa* (Volkova 1969) Downie 1982 – Zang, pp. 104–105, Pl. 2:F–H. □1992 *Skiagia ciliosa* (Volkova) Downie 1982 – Zang & Walter, p. 99, Fig. 20A–E. □1993 *Skiagia ciliosa* (Volkova 1969) Downie 1982 – Vidal & Peel, p. 29, Fig. 12a, d. □*non* 1993 *Skiagia ciliosa* (Volkova 1969) Downie 1982 – Vidal & Peel, Fig. 12b, c. □1995 *Skiagia ciliosa* (Volkova 1968) Downie 1982 – Vidal *et al.*, Fig. 7:3–4.

Material. – Sixty-one well-preserved and satisfactorily preserved specimens.

Dimensions. – *N* = 17. Vesicle diameter 23–39 μm, length of processes 3–8 μm.

Present record. – The Upper Silesia area, Goczałkowice IG-1 borehole, Goczałkowice formation at a depth of 2854.4–2973.7 m, Lower Cambrian, *Holmia kjerulfi* Zone; at a depth of 2771.8–2791.4 m, Lower Cambrian, probably *Protolenus* Zone; and at a depth of 2766.8–2771.1 m, Middle Cambrian, *Acadoparadoxides oelandicus* Superzone; Sosnowiec IG-1 borehole, Sosnowiec Formation at a depth of 3403.5–3407.2 m; Middle Cambrian, *Acadoparadoxides oelandicus* Superzone.

Occurrence and stratigraphic range. – The species has been recorded in the Lower Cambrian, *Holmia kjerulfi* and *Protolenus* Zones, and in the Middle Cambrian, *Acadoparadoxides oelandicus* Superzone and their time-equivalents in Poland, Latvia, Lithuania, Estonia, the Ukraine, Sweden, Denmark, Norway, England, Wales, Svalbard, Greenland and Kazakhstan (see Moczydłowska 1991, pp. 66–67). Additionally, it has been recorded from Sweden, Island of Gotland, Grötlingbo-1 borehole and Gotska Sandön Island, Gotska Sandön borehole, File Haidar Formation and 'oelandicus beds', Gulf of Bothnia, Finngrun-

det borehole, Söderfjärden Formation, and Finland, Vaasa area, Söderfjärden-3 borehole, Söderfjärden Formation, Lower Cambrian, *Holmia kjerulfi* and *Protolenus* Zones, and Middle Cambrian, *Acadoparadoxides oelandicus* Superzone; south-central Sweden, Närke area at Kvarntorp, 'oelandicus beds', Middle Cambrian, *Acadoparadoxides oelandicus* Superzone (Hagenfeldt 1989a). Spain, the Iberian Mountains, Ribota Formation and Daroca Sandstones, Lower Cambrian (Gamez *et al.* 1991), time equivalent to the *Holmia kjerulfi* and *Protolenus* Zones, respectively. Central Australia, the Amadeus Basin, Hermannsburg 41 borehole, Tempe Formation, lowermost Middle Cambrian (Zang & Walter 1992). South China, Yunnan Province, Chengjiang County at Qiongzhusi and Maotianshan borehole, Qiongzhusi Formation, Lower Cambrian, *Eoredlichia–Wutingaspis* Zone (Zang 1992). North Greenland, Peary Land, Buen Formation, Lower Cambrian, time equivalent to the *Holmia kjerulfi* and *Protolenus* Zones (Vidal & Peel 1993). Eastern Siberia, Yakutia, the Kharaulakh Mountains at Chekurovka, Tyusersk Formation, Lower Cambrian, Tommotian Stage, *Dokidocyathus regularis* Zone (Vidal *et al.* 1995).

Skiagia compressa (Volkova, 1968) Downie, 1982

Fig. 39C

Synonymy and description. – See Moczydłowska 1991, p. 67. Additionally: □1989 *Skiagia compressa* (Volkova) Downie 1982 – Hagenfeldt 1989a, Pl. 5:2. □1993 *Skiagia compressa* (Volkova 1968) Downie 1982 – Vidal & Peel, p. 29, Fig. 14g. □1992 *Skiagia compressa* (Volkova 1968) sensu Downie 1982 – Zang, pp. 105–106, Pl. 2L–N. □1992 *Skiagia* sp. – Zang, Pl. 2I.

Material. – Four poorly preserved specimens.

Dimensions. – *N* = 4. Vesicle diameter 28–36 μm, length of processes 5–10 μm.

Present record. – The Upper Silesia area, Goczałkowice IG-1 borehole, Goczałkowice formation at a depth of 2966.0–2969.2 m, Lower Cambrian *Holmia kjerulfi* Zone, and at a depth of 2766.8–2771.1 m, Middle Cambrian, *Acadoparadoxides oelandicus* Superzone.

Occurrence and stratigraphic range. – The species has been recorded in Poland, Estonia, Latvia, Lithuania, Russia, Ukraine, Sweden, Norway, Scotland, Belgium, Svalbard, North and East Greenland in the Lower Cambrian strata, time equivalent to the *Holmia kjerulfi* and *Protolenus* Zones (compilation *in* Moczydłowska 1991, p. 67). In Poland it has also been recorded in the Middle Cambrian *Acadoparadoxides oelandicus* Superzone (Volkova *et al.* 1979, Moczydłowska 1991). Additionally: Sweden, Island

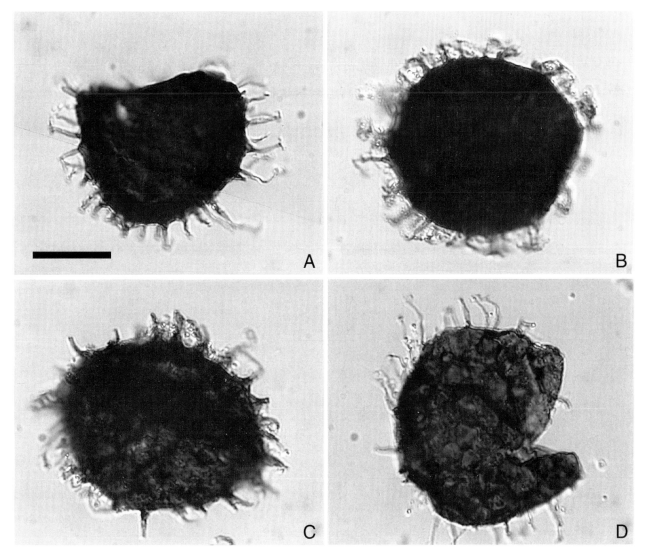

Fig. 39. □A–B. *Skiagia ciliosa* (Volkova) Downie. A, PMU-Pl.123-Z/37; B, PMU-Pl.124-K/31/1, showing funnel-like tips of processes and conical bases. □C. *Skiagia compressa* (Volkova) Downie. PMU-Pl.125-T/39/1, characteristic wavy outline of the vesicle. □D. *Skiagia orbiculare* (Volkova) Downie. PMU-Pl.126-P/40/3, with thin, tubular processes and small funnel tips. The Goczałkowice IG-1 borehole, depth 2766.8–2771.1 m, Goczałkowice formation, Middle Cambrian, *Acadoparadoxides oelandicus* Superzone. Scale bar in A equals 12 μm for all micrographs.

of Gotland, Grötlingbo-1 borehole and Gotska Sandön Island, Gotska Sandön borehole, File Haidar Formation and 'oelandicus beds', Gulf of Bothnia, Finngrundet borehole, Söderfjärden Formation; and Finland, Vaasa area, Söderfjärden-3 borehole, Söderfjärden Formation; Lower Cambrian, *Holmia kjerulfi* and *Protolenus* Zones, and Middle Cambrian, *Acadoparadoxides oelandicus* Superzone (Hagenfeldt 1989a). South China, Yunnan Province, Chengjiang County at Qiongzhusi and Maotianshan borehole, Qiongzhusi Formation, Lower Cambrian, *Eoredlichia–Wutingaspis* Zone (Zang 1992). North Greenland, Peary Land, Buen Formation, Lower Cambrian, time equivalent to the *Holmia kjerulfi* and *Protolenus* Zones (Vidal & Peel 1993).

Skiagia orbiculare (Volkova, 1968) Downie, 1982

Fig. 39D

Synonymy and description. – See Moczydłowska 1991, p. 68. Additionally: □1992 *Skiagia orbiculare* (Volkova 1968) Downie 1982 – Zang, p. 106, Pl. 2J–K. □1993 *Skiagia orbiculare* (Volkova, 1968) Downie, 1982 – Vidal & Peel, p. 31, Fig. 13d, e. □1993 *Skiagia ciliosa* (Volkova, 1969) Downie, 1982 (morphotype A) – Vidal & Peel, Fig. 12c. □1995 *Skiagia orbiculare* (Volkova 1968) Downie 1982 – Vidal *et al.*, Fig. 7:6.

Material. – Eight poorly preserved specimens.

Dimensions. – $N=7$. Vesicle diameter 27–36 μm, length of processes 5–8 μm.

Present record. – The Upper Silesia area, Goczałkowice IG-1 borehole, Goczałkowice formation at depths of 2854.4–2862.0 m and 2771.8–2789.0 m, Lower Cambrian, *Holmia kjerulfi* Zone and probably *Protolenus* Zone, respectively; at a depth of 2766.8–2771.1 m, Middle Cambrian, *Acadoparadoxides oelandicus* Superzone. The present record is the first in Middle Cambrian strata.

Occurrence and stratigraphic range. – The species has a wide palaeogeographic distribution and has been recorded in Poland, Estonia, Latvia, Ukraine, Sweden, Norway, Svalbard, North and East Greenland, and Kazakhstan in Lower Cambrian strata, time equivalent to the *Schmidtiellus mickwitzi*, *Holmia kjerulfi* and *Protolenus* Zones (see compilation *in* Moczydłowska 1991, p. 68). The new record is from Sweden, Island of Gotland, Grötlingbo-1 borehole and Gotska Sandön Island, Gotska Sandön borehole, File Haidar Formation, and Finland, Vaasa area, Söderfjärden-3 borehole, Söderfjärden Formation, Lower Cambrian, *Holmia kjerulfi* and *Protolenus* Zones (Hagenfeldt 1989a). South China, Yunnan Province, Chengjiang County at Qiongzhusi and Maotianshan borehole, Qiongzhusi Formation, Lower Cambrian, *Eoredlichia–Wutingaspis* Zone (Zang 1992). North Greenland, Peary Land, Buen Formation, Lower Cambrian, time equivalent to the *Holmia kjerulfi* and *Protolenus* Zones (Vidal & Peel 1993). Eastern Siberia, Yakutia, the Kharaulakh Mountains at Chekurovka, Tyusersk Formation, Lower Cambrian, Tommotian Stage, *Dokidocyathus regularis* Zone (Vidal *et al.* 1995).

Skiagia ornata (Volkova, 1968) Downie, 1982

Fig. 40A–D

Synonymy and description. – See Moczydłowska 1991, pp. 68–69. Additionally: □1989 *Skiagia ornata* (Volkova) Downie, 1982 – Hagenfeldt 1989a, pp. 121–123, Pl. 5:6. □1992 *Skiagia ornata* (Volkova 1968) sensu Downie 1982 – Zang, pp. 106–107, Pl. 1O–Q. □1993 *Skiagia ciliosa* (Volkova, 1969) Downie, 1982 (morphotype B) – Vidal & Peel, Fig. 12b. □1993 *Skiagia ornata* (Volkova, 1968) Downie, 1982 – Vidal & Peel, p. 31, Fig. 13a. □1995 *Skiagia ornata* (?) (Volkova 1968) Downie 1982 – Vidal *et al.*, Fig. 7:5.

Material. – Sixty well-preserved and satisfactorily preserved specimens.

Dimensions. – $N=18$. Vesicle diameter 23–54 m, length of processes 8–19 μm.

Present record. – The Upper Silesia area, Goczałkowice IG-1 borehole, Goczałkowice formation at depths of 2952.0–2973.7 m and 2771.8–2789.0 m, Lower Cambrian, *Holmia kjerulfi* Zone and probably *Protolenus* Zone, respectively; at a depth of 2766.8–2771.1 m, Middle Cambrian *Acadoparadoxides oelandicus* Superzone. This is the first record in Middle Cambrian strata.

Occurrence and stratigraphic range. – The species has been recorded in Poland, Estonia, Latvia, Ukraine, Sweden, Norway, Denmark, Svalbard, North and East Greenland, and Kazakhstan in Lower Cambrian strata, time equivalent to the *Schmidtiellus mickwitzi*, *Holmia kjerulfi* and *Protolenus* Zones (see compilation *in* Moczydłowska 1991, p. 69). Additional occurrences are in Sweden, Gotska Sandön Island, Gotska Sandön borehole, File Haidar Formation, and Gulf of Bothnia, Finngrundet borehole, Söderfjärden Formation, Lower Cambrian, *Holmia kjerulfi* and *Protolenus* Zones (Hagenfeldt 1989a). South China, Yunnan Province, Chengjiang County at Qiongzhusi and Maotianshan borehole, Qiongzhusi Formation, Lower Cambrian, *Eoredlichia–Wutingaspis* Zone (Zang 1992). North Greenland, Peary Land, Buen Formation, Lower Cambrian, time equivalent to the *Holmia kjerulfi* and *Protolenus* Zones (Vidal & Peel 1993). Eastern Siberia, Yakutia, the Kharaulakh Mountains at Chekurovka, Tyusersk Formation, Lower Cambrian, Tommotian Stage, *Dokidocyathus regularis* Zone (Vidal *et al.* 1995).

Skiagia sp.

Material. – Twenty-eight poorly preserved specimens.

Present record. – The Upper Silesia area, Goczałkowice IG-1 borehole, Goczałkowice formation in the interval of 2766.8–2973.7 m, Lower Cambrian, *Holmia kjerulfi* Zone, to Middle Cambrian, *Acadoparadoxides oelandicus* Superzone; Sosnowiec IG- borehole, Sosnowiec formation at a depth of 3365.2–3372.3 m, Middle Cambrian, *Acadoparadoxides oelandicus* Superzone.

Genus *Solisphaeridium* Staplin *et al.*, 1965, emended

Type species. – *Solisphaeridium stimuliferum* (Deflandre, 1938) Staplin *et al.*, 1965 (=*Micrhystridium (Hystrichosphaeridium) stimuliferum* Deflandre, 1938, p. 192, Pl. 10:10); France, Upper Jurassic (Staplin *et al.* 1965, p. 183; Sarjeant 1968, p. 223).

Emended diagnosis. – Organic-walled microfossils with vesicles of medium size, circular to oval in outline, originally spherical, having a single-layered firm wall with psi-

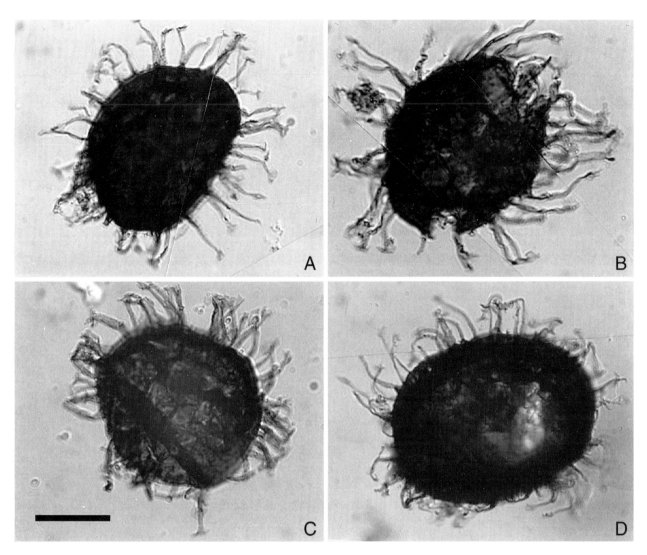

Fig. 40. Skiagia ornata (Volkova) Downie. □A. PMU-Pl.127-G/26/4. □B. PMU-Pl.128-R/43/1. □C. PMU-Pl.129-Y/25/4. □D. PMU-Pl.130-V/46/3. Specimens showing long, prominent, tubular processes with funnel tips and sharp, regular outline of vesicle. The Goczałkowice IG-1 borehole, depth 2766.8–2771.1 m, Goczałkowice formation, Middle Cambrian, *Acadoparadoxides oelandicus* Superzone. Scale bar in C equals 12 μm for all micrographs.

late to granulate surface and bearing processes that are hollow and communicate with the cavity of the vesicle. The processes are of variable length, simple, cylindrical or tapering towards the distal terminations, with widened bases or narrow bases that arise perpendicular to the vesicle wall. The tips are sharp-pointed, rounded or blunt, occasionally divided. Excystment is by median splitting.

Remarks. – In the original diagnosis of *Solisphaeridium* (Staplin *et al.* 1965, pp. 183–184) it was stated that the processes are hollow or solid. The proposed emendation considers the uniformly hollow nature of the processes.

The genus differs from *Heliosphaeridium* Moczydłowska, 1991, in its greater dimensions, its overall size and diameter of the vesicle, and in having more robust and rigid processes. It differs from *Baltisphaeridium* Eisenack,

1958, emend. Eisenack, 1969, in having free communication between the cavities of processes and vesicle, and from *Multiplicisphaeridium* Staplin, 1961, emend. Eisenack, 1969, in having simple processes.

Solisphaeridium baltoscandium Eklund, 1990 emended

Fig. 40A

Synonymy. – □1979 *Baltisphaeridium* sp. 2. – Volkova *et al.*, p. 13, Pl. 6:10–12. □1990 *Solisphaeridium baltoscandium* sp. nov – Eklund, p. 41, Fig. 8:I. □1994 *Polygonium? baltoscandium* (Eklund 1990, p. 41, Fig. 8:I) comb.nov. – Sarjeant & Stancliffe, p. 43.

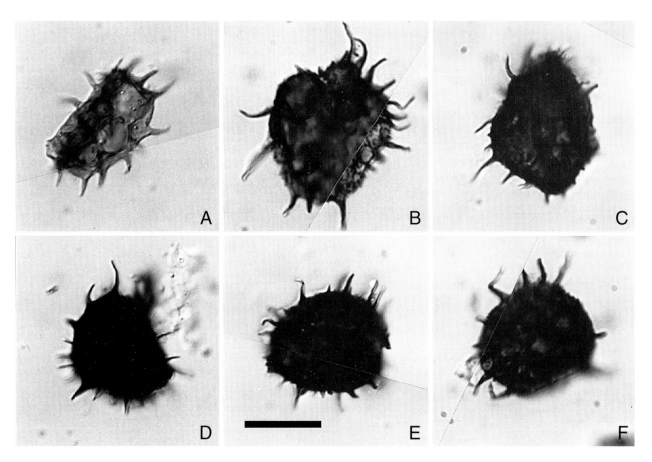

Fig. 41. □A. *Solisphaeridium baltoscandium* Eklund, emended. PMU-Pl.131-K/35/3, with clearly visible hollow processes communicating with inner vesicle cavity. □B–F. *Solisphaeridium bimodulentum* n.sp.; B, PMU-Pl.132-G/34/1, holotype, displaying processes of bimodal length and granular vesicle wall sculpture; C, PMU-Pl.133-F/30/1; D, PMU-Pl.134-Q/32/3, probable circular opening (pylome) in right part of vesicle; E, PMU-Pl.135-D/29/3, two types of processes: short, thorn-like and long, conical ones; F, PMU-Pl.136-K/46, denticulate outline of the vesicle. The Sosnowiec IG-1 borehole, Sosnowiec formation. A–C, E, depth 3403.5–3407.2 m, and F, depth 3365.2–3372.3 m, Middle Cambrian, *Acadoparadoxides oelandicus* Superzone; D, depth 3174.0–3174.7 m, lower Upper Cambrian. Scale bar in E equals 12 µm for all micrographs.

Material. – Seven well-preserved specimens.

Emended diagnosis. – Vesicles circular to oval in outline, originally spherical, bearing sparse rigid conical processes. The basal parts of the processes may form an undulating outline of the vesicle. Distally the processes taper to sharp-pointed tips. Occasionally, individual processes may have bifurcated tips. The processes are hollow and connected with the inner cavity of the vesicle.

Dimensions. – $N=5$. Diameter of vesicle 16–25 µm, length of processes 4–6 µm.

Remarks. – The diagnosis of *Solisphaeridium baltoscandium* (Eklund 1990, p. 41) lacks information on whether the processes are hollow and connected with the inner cavity of the vesicle. These features are observed on the micrograph of the holotype of the species (Eklund 1990,

Fig. 8I) and are stated in the description of the synonymous species *Baltisphaeridium* sp. 2 (Volkova *et al.* 1979, p. 13; Eklund 1990, p. 41). The character of the processes and their connection with the vesicle cavity is regarded as a diagnostic feature, and it is therefore incorporated into the emended diagnosis of *S. baltoscandium*.

Present record. – The Upper Silesia area, Sosnowiec IG-1 borehole, Sosnowiec formation at a depth of 3403.5–3407.2 m; Middle Cambrian, *Acadoparadoxides oelandicus* Superzone.

Occurrence and stratigraphic range. – Latvia, Lower Cambrian, Vergale horizon (Volkova *et al.* 1979). Sweden, Östergötland, File Haidar Formation (Lingulid Sandstone member), Lower Cambrian, Rausve 'stage' (Eklund 1990), i.e. time equivalent to the *Holmia kjerulfi* and *Protolenus* Zones (Moczydłowska 1991).

Solisphaeridium bimodulentum n.sp.

Fig. 41B–F

Holotype. – Specimen PMU-Pl.132-G/34/1; Fig. 41B.

Derivation of name. – From Latin *bis* – twice; *modus* – mode, ratio, proper measure, norm; and *-lentus* – full of; referring to the bimodal length of the processes.

Locus typicus. – The Upper Silesia area, Sosnowiec IG-1 borehole (Figs. 2 and 3).

Stratum typicum. – Mudstones in the Sosnowiec formation at a depth of 3403.5–3407.2 m; Middle Cambrian, *Acadoparadoxides oelandicus* Superzone.

Material. – Eighteen fairly well-preserved specimens.

Diagnosis. – Vesicles circular to oval in outline, originally spherical, having a dense wall and bearing robust processes. The surface of the vesicle is ornamented by small granae shaped as delicate denticles on the outline of the vesicle. The processes are hollow, and their cavities are freely connected with the inner cavity of the vesicle. There are two types of processes: longer and more numerous (approximately 12–15 observed on the vesicle contour) with conical bases, tapering towards sharp-pointed tips, and shorter, thorn-like processes (6–8 observed on the vesicle contour) that are distributed between the longer processes.

Dimensions. – $N=8$. Diameter of vesicle 17–27 μm (holotype 20x23 μm), longer processes 5–7 μm long; shorter processes 2–3 μm. Dimensions of processes show bimodal distribution of length.

Remarks. – Specimens illustrated in Fig. 41D, E are light brown because of the thermal alteration which resulted in the apparent opaque appearance of the specimens in micrographs produced through interference contrast microscopy. Originally, the vesicle wall of the microfossils is translucent, the processes are clearly hollow and their cavities are connected with the vesicle cavity (Fig. 41 B).

Present record. – The Upper Silesia area, Sosnowiec IG-1 borehole, Sosnowiec formation at depths of 3403.5–3407.2 m and 3365.2–3372.3 m, and 3174.0–3174.7 m; Middle Cambrian, *Acadoparadoxides oelandicus* Superzone and lower Upper Cambrian, respectively.

Solisphaeridium cylindratum n.sp.

Fig. 42A–D

Holotype. – Specimen PMU-Pl.137-Z/33/3; Fig. 42A. Paratype specimen PMU-Pl.139-J/44/4; Fig. 42C.

Derivation of name. – From Latin *cylindratus* – cylindrical, referring to the shape of the processes.

Locus typicus. – The Upper Silesia area, Sosnowiec IG-1 borehole (Figs. 2 and 3).

Stratum typicum. – Mudstones in the Sosnowiec formation at a depth of 3403.5–3407.2 m; Middle Cambrian, *Acadoparadoxides oelandicus* Superzone.

Material. – Thirty-six well-preserved specimens.

Diagnosis. – Vesicles circular to oval in outline, originally spherical, bearing not very numerous (approximately 16–20 observed on the vesicle contour), narrow, cylindrical processes. The proximal part of the processes is slightly widened at the junction with the vesicle wall, but some processes remain cylindrical throughout their entire length. The tips of the processes are rounded. The processes are hollow and communicate with the cavity of the vesicle.

Dimensions. – $N=17$. Diameter of vesicle 16–27 μm (holotype 18×25 μm), length of processes 3–7 μm (holotype 7 μm).

Remarks. – *S. cylindratum* n.sp. differs from other species of *Solisphaeridium* in having cylindrical processes with rounded tips. Specimens that are thermally altered appear to be opaque (Fig. 42D) and their processes may appear solid, a feature that is an artefact of preservation and partly a result of enhanced contrast depending on the use of interference contrast microscopy to reveal the morphology of the processes.

Present record. – The Upper Silesia area, Sosnowiec IG-1 borehole, Sosnowiec formation at depths of 3403.5–3407.2 m and 3365.2–3372.3 m; Middle Cambrian, *Acadoparadoxides oelandicus* Superzone.

Solisphaeridium elegans n.sp.

Fig. 42E–F

Holotype. – Specimen PMU-Pl.141-W/45; Fig. 42E.

Synonymy. – ☐1982 *Goniosphaeridium implicatum* (Fridrichsone) comb.nov. – Downie, p. 278, Fig. 10v–x.

Derivation of name. – From Latin *elegans* – fine, select, referring to the overall, very shapely appearance of the vesicle.

Locus typicus. – The Upper Silesia area, Sosnowiec IG-1 borehole (Figs. 2 and 3).

Stratum typicum. – Mudstones of the Sosnowiec formation at a depth of 3403.5–3407.2 m; Middle Cambrian, *Acadoparadoxides oelandicus* Superzone.

Material. – Four well-preserved specimens.

Diagnosis. – Vesicles circular to oval in outline, originally spherical, with a smooth wall, bearing numerous

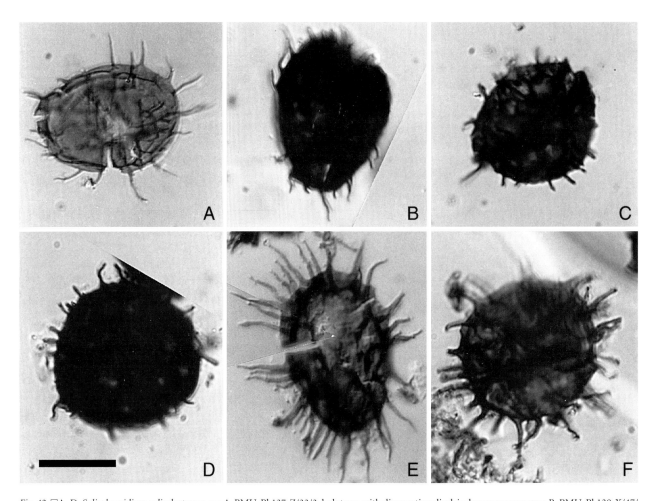

Fig. 42. □A–D. *Solisphaeridium cylindratum* n.sp.; A, PMU-Pl.137-Z/33/3, holotype, with diagnostic cylindrical narrow processes; B, PMU-Pl.138-X/47/ 3; C, PMU-Pl.139-I/44/4, paratype; D, PMU-Pl.140-H/37/1. □E–F. *Solisphaeridium elegans* n.sp.; E, PMU-Pl.141-W/45, holotype, showing prominent processes, long, regular in shape and tapering gradually; F, PMU-Pl.142-E/29/2. The Sosnowiec IG-1 borehole, Sosnowiec formation. A, E, F, depth 3403.5–3407.2 m, and B–D, depth 3365.2–3372.3 m, Middle Cambrian, *Acadoparadoxides oelandicus* Superzone. Scale bar in D equals 12 μm for all micrographs.

(approximately 25–34 observed on the vesicle contour) long, tubular processes that are hollow and open into the vesicle cavity. The processes are widened proximally, having transitional junctions with the vesicle wall, and taper distally. The tips of the processes are sharp or truncated. Discrete processes may have bifurcated tips.

Dimensions. – *N*=4. Diameter of vesicle 18–25 μm (holotype 18×25μm), length of processes 7–11 μm (holotype 9–11 μm).

Present record. – As for the holotype.

Occurrence and stratigraphic range. – Scotland, Knockan and Skiag Bridge, Fucoid Beds, Lower Cambrian, equivalent to *Holmia kjerulfi* Zone (Downie 1982).

Solisphaeridium flexipilosum Slaviková, 1968, emended

Fig. 43A–D

Synonymy. – □1968 *Solisphaeridium flexipilosum* n.sp. – Slaviková, p. 200, Pl. 1:6, 8. □1976 *Acanthodiacrodium* sp. – Vavrdová, Pl. 1:2. □1987 *Solisphaeridium flexipilosum* Slaviková, 1968 – Fombella, Pl. 2:24. □1994 *Micrhystridium flexipilosum* (Slaviková 1968, p. 200, Pl. 1:6, 9) comb.nov. – Sarjeant & Stancliffe, p. 16.

Material. – Twelve very well-preserved specimens.

Emended diagnosis. – Vesicles circular to oval in outline, originally spherical, with a smooth wall, bearing numer-

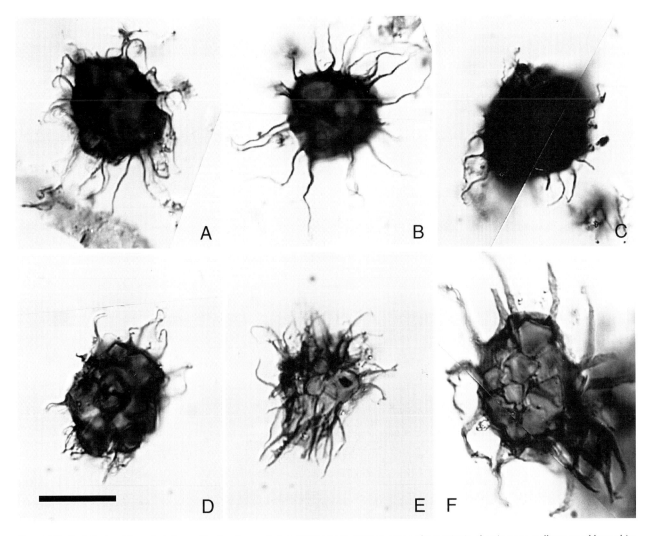

Fig. 43. □A–D. *Solisphaeridium flexipilosum* Slaviková emended. A, PMU-Pl.143-K/29; B, PMU-Pl.144-E/34/3, showing very well preserved long, thin, simple and flexible processes; C, PMU-Pl.145-I/31; D, PMU-Pl.146-P/45/4; E, *Solisphaeridium multiflexipilosum* Slaviková emended. PMU-Pl.147-Z/35/3. □F. *Solisphaeridium* sp. A. PMU-Pl.148-N/34/3, clearly visible, wide and hollow processes, freely communicating with vesicle cavity. The Sosnowiec IG-1 borehole, depth 3174.0–3174.7 m, Sosnowiec formation, lower Upper Cambrian. Scale bar in D equals 12 μm for all micrographs.

ous (approximately 15–18 observed on the vesicle contour) long and flexible processes that are hollow and connected with the interior of the vesicle. The processes are slim, slightly widened at the base and taper towards the distal terminations. The process tips are sharp-pointed.

Dimensions. – *N*=8. Diameter of vesicle 19–25 μm, length of processes 8–14 μm.

Remarks. – The free communication between the cavities of processes and vesicle is herein added as a diagnostic feature of the species to the original diagnosis by Slaviková (1968).

Present record. – The Upper Silesia area, Sosnowiec IG-1 borehole, Sosnowiec formation at a depth of 3203.5–3205.5 m, Middle Cambrian, probably *Paradoxides forch-*

hammeri Superzone; and at a depth of 3174.0–3174.7 m, lower Upper Cambrian.

Occurrence and stratigraphic range. – Czech Republic, Bohemian Massif at Strašice, Těně Sš-III borehole, Jince Formation, Middle Cambrian, *Ellipsocephalus hoffi* Subzone (Slaviková 1968), and Jince, Jince Formation, Middle Cambrian, *Eccaparadoxides pusillus, Paradoxides gracilis* and *Hydrocephalus lyelli* Zones (Vavrdová 1976). The Middle Cambrian successions in the Bohemian Massif are correlated with the *Paradoxides paradoxissimus* Superzone, without the *Ptychagnostus gibbus* Zone (Öpik 1979). Spain, the Cantabrian Mountains at Vozmediano, interpreted by Fombella (1987) as Upper Cambrian but here referred to Middle Cambrian (Aramburu *et al.* 1992, see under 'Stratigraphic ranges').

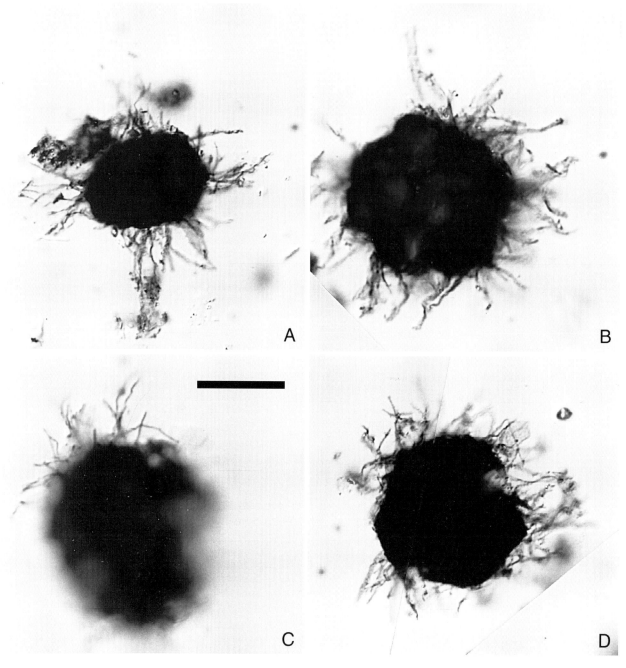

Fig. 44. □A–D. *Solisphaeridium implicatum* (Fridrichsone) n.comb.; A, PMU-Pl.149-Q/46; B, PMU-Pl.150-Z/35/2; C, PMU-Pl.151-H/31; D, PMU-Pl.152-Q/43. All specimens displaying dense opaque central body and translucent, long and slender processes. The Sosnowiec IG-1 borehole, Sosnowiec formation. A, B, D, depth 3174.0–3174.7m, lower Upper Cambrian; C, depth 3403.5–3407.2 m, Middle Cambrian, *Acadoparadoxides oelandicus* Superzone. Scale bar in C equals 18 µm for A and D; 12 µm for B and C.

Solisphaeridium implicatum (Fridrichsone, 1971) n.comb.

Fig. 44A–D

Synonymy.. – □1969 *Baltisphaeridium* sp. 1 – Volkova 1969b, p. 226, Pl. 49:20. □1971 *Baltisphaeridium implicatum* Fridrichsone sp.nov. – Fridrichsone, pp. 11–12, Pl.

3:7–14. □*nomen nudum* 1979 *Comasphaeridium piliferum* n.sp. – Fombella, Pls. 4:66; 5:80. □*non* 1982 *Goniosphaeridium implicatum* (Fridrichsone 1971) comb.nov. – Downie, p. 278, Fig. 10v–x. □1987 *Baltisphaeridium implicatum* Fridrichsone, 1971 – Knoll & Swett, p. 915, Fig. 7.7. □1989 *Baltisphaeridium implicatum* Fridrichsone, 1971 – Hagenfeldt, 1989a, pp. 21–24, Pl. 1:5. □1991

Goniosphaeridium implicatum (Fridrichsone, 1971) Downie, 1982 – Moczydłowska, p. 55, Pl. 10A. □*non* 1992 *Goniosphaeridium* cf. *implicatum* (Fridrichsone) sensu Downie 1982 – Zang, p. 101, Pl. 3C–D. (Note misprint on p. 101, as Pl. 2C–D). □1994 *Polygonium implicatum* (Fridrichsone 1971, p. 11–12, pl. 3, figs 7–14) comb.nov. – Sarjeant & Stancliffe, p. 43.

Material. – Seventy-four specimens in variable states of preservation. .

Description. – Vesicles circular to irregular in outline, originally spherical, consisting of an opaque central body and very numerous (more than 30 observed on the vesicle contour), translucent, long and slender processes. The processes are tubular and narrow having widened bases and sharp-pointed or blunt tips. The cavity of the processes communicates freely with the inner cavity of the vesicle.

Dimensions. – N=17. Diameter of central body 18–36 µm, length of processes 7–15 µm.

Remarks. – The new taxonomic combination is proposed for the species originally described as *Baltisphaeridium implicatum* (Fridrichsone 1971) and then transferred to *Goniosphaeridium implicatum* (Downie 1982). However, the specimens attributed by Downie (1982, Fig. 10v–x) to *G. implicatum* are not conspecific with *Baltisphaeridium implicatum* Fridrichsone, 1971 (Moczydłowska 1991, p. 55). These specimens belong to a new species and are here regarded as conspecific with *Solisphaeridium elegans* n.sp. The transfer of *B. implicatum* to the genus *Goniosphaeridium* was followed by Moczydłowska (1991) and Moczydłowska & Crimes (1995), who assumed that its processes are hollow and communicating with the vesicle cavity, though this feature was not stated in the original diagnosis by Fridrichsone (1971). The specimen of *B. implicatum* described under an open nomenclature as *Baltisphaeridium* sp. 1 and regarded as a synonym of *B. implicatum* (Volkova 1969b) has clearly hollow processes, but their connections with the vesicle are obscured by the opaque central body and the poor state of preservation (Volkova 1969b, p. 226).

The bases of the processes in *B. implicatum* are not well developed and not as wide as diagnosed for *Goniosphaeridium* Eisenack, 1969, emend. Kjellström, 1971. Their morphology is more comparable to the processes of *Solisphaeridium*. Additionally, the taxonomic status of *Goniosphaeridum* has been re-evaluated, and the genus is regarded as a junior synonym of *Polygonium* Vavrdová, 1966 (Turner 1984; Albani 1989; Le Hérissé 1989; Fensome *et al.* 1990).

The conical bases of the processes and free communication between the cavities of processes and central body is characteristic for two genera: *Solisphaeridium* and *Polygonium*. The genus *Polygonium* has a distinctive polygonal outline of the vesicle that is formed by exceptionally wide process bases, whereas *Solisphaeridium* has a circular to oval outline and narrow process bases.

Present record. – The Upper Silesia area, Goczałkowice IG-1 borehole, Goczałkowice formation at a depth of 2766.8–2771.1 m, Middle Cambrian, *Acadoparadoxides oelandicus* Superzone; Sosnowiec IG-1 borehole, Sosnowiec formation in the interval of 3365.2–3423.8 m, Middle Cambrian, *Acadoparadoxides oelandicus* Superzone; and at a depth of 3174.0–3174.7 m, lower Upper Cambrian.

Occurrence and stratigraphic range. – Poland, Lublin Slope of the East European Platform, Kaplonosy and Radzyń formations, Lower Cambrian, *Holmia kjerulfi* and *Protolenus* Zones, and Kostrzyń Formation, Middle Cambrian, *Acadoparadoxides oelandicus* Zone (Volkova 1969a, b; Moczydłowska 1991). Latvia, Lower Cambrian, Vergale and Rausve horizons and Middle Cambrian, Kibartai horizon (Fridrichsone 1971; Volkova *et al.* 1979). The Ukraine, Lower Cambrian, Vergale and Rausve horizons (Volkova *et al.* 1979). Spain, the Cantabrian Mountains at Vozmediano, the Oville Formation attributed to as Upper Cambrian by Fombella (1979) but here considered to be Middle Cambrian (see under 'Stratigraphic ranges'). Sweden, Västergötland, File Haidar Formation, Lower Cambrian, equivalent to *Holmia kjerulfi* and *Protolenus* Zones (Moczydłowska & Vidal 1986; Moczydłowska 1991); Närke area at Kvarntorp, Island of Gotland, Grötlingbo-1 borehole and Gotska Sandön Island, upper File Haidar Formation and 'oelandicus' beds, upper Lower Cambrian and Middle Cambrian, *Acadoparadoxides oelandicus* Superzone; Gulf of Bothnia, Finngrundet borehole, upper Söderfjärden Formation, Middle Cambrian, *Acadoparadoxides oelandicus* Superzone (Hagenfeldt 1989a). Finland, Vaasa area, Söderfjärden borehole, Söderfjärden Formation, upper Lower Cambrian and Middle Cambrian, *Acadoparadoxides oelandicus* Superzone (Hagenfeldt 1989a). Svalbard, East Spitsbergen, South Tokammane, Topiggane and Andromedafjellet, Tokammane Formation, Lower Cambrian (Knoll & Swett 1987).

Solisphaeridium multiflexipilosum Slaviková, 1968, emended

Fig. 43 E

Synonymy. – □1968 *Solisphaeridium multiflexipilosum* n.sp. – Slaviková, p. 200, Pl. 1:4–5. □1994 *Comasphaeridium*? *multiflexipilosum* (Slaviková 1968, p. 200, Pl. 1:4–5) comb.nov. – Sarjeant & Stancliffe, p. 27.

Material. – Two satisfactorily preserved specimens.

Emended diagnosis. – Organic-walled microfossils with vesicles circular to oval in outline, originally spherical, having psilate wall and bearing very numerous (approximately 30 observed on the vesicle contour), long and flexible processes. The processes are tubular, wider at the base and tapering distally to sharp-pointed or blunt tips. Occasionally, some tips are slightly widened, funnel-like. The processes are hollow and connected with the inner cavity of the vesicle.

Dimensions. – $N=2$. Diameter of central body 18–20 µm, length of processes 7–11 µm.

Remarks. – The free communication between the cavities of processes and vesicle, not mentioned by Slaviková (1968), is here included in the emended diagnosis of the species.

Present record. – The Upper Silesia area, Sosnowiec IG-1 borehole, Sosnowiec formation at a depth of 3174.0–3174.7 m; lower Upper Cambrian.

Occurrence and stratigraphic range. – Czech Republic, Bohemian Massif, Strasice, Tene Ss-III borehole, Jince Formation, Middle Cambrian, *Ellipsocephalus hoffi* subzone (Slaviková 1968); Jince and Stryje, Jince Formation, Middle Cambrian, *Eccaparadoxides pusillus, Paradoxides gracilis* and *Hydrocephalus lyelli* Zones (Vavrdová 1976). According to the correlation by Öpik (1979), these strata are time-equivalent to the *Paradoxides paradoxissimus* Superzone excluding the *Ptychagnostus gibbus* Zone.

Solisphaeridium sp. A

Fig. 43F

Material. – One poorly preserved specimen with distinctive, thick, tubular processes tapering to acuminate tips and having only slightly widened bases.

Dimensions. – Vesicle diameter 18–23 µm, process length 10–16 µm.

Present record. – The Upper Silesia area, Sosnowiec IG-1 borehole, Sosnowiec formation at a depth of 3174.0–3174.7 m, lower Upper Cambrian.

Solisphaeridium sp.

Material. – Three poorly preserved specimens.

Present record. – The Upper Silesia area, Goczałkowice IG-1 borehole, Goczałkowice formation at a depth of 2766.8–2771.1 m, Middle Cambrian, *Acadoparadoxides oelandicus* Superzone.

Genus *Stelliferidium* Deunff, Górka & Rauscher, 1974

Type species. – *Stelliferidium striatulum* (Vavrdová, 1966) Deunff, Górka & Rauscher, 1974; Vavrdová 1966, pp. 411–412, Pl. 2:3 (holotype), Pl. 1:2 (Deunff *et al.* 1974, p. 16), Czech Republic, the Bohemian Massif at Klabava, the Klabava Shales, Arenig.

Stelliferidium robustum n.sp.

Fig. 45A, B, D

Synonymy. – □1986 *Stelliferidium* sp. A – Welsch, pp. 82–83, Pl. 4:9–13.

Holotype. – Specimen PMU-Pl.153-R/39; Fig. 45A. Paratype specimen PMU.Pl.154-O/36/4; Fig. 45B.

Derivation of name. – From Latin *robustus* (adjective), robust, firm, referring to the appearance of processes.

Locus typicus. – The Upper Silesia area, Sosnowiec IG-1 borehole (Figs. 2 and 3).

Stratum typicum. – A mudstone interbed in fine-grained sandstones of the Sosnowiec formation at a depth of 3166.3–3167.7 m, Upper Cambrian.

Diagnosis. – Vesicle originally spherical, circular in outline, bearing evenly distributed, numerous, robust processes. The processes are wide, cylindrical with slightly conical bases and divided or ramifying distal portions, forming pinnulae of second and third order. The corona of pinnulae is small in relation to the the dimensions of the whole process. The process bases have very dense, opaque portions (plugs?) at the transition towards the central cylindrical portion of the process. Occasionally, the bases gradually taper into the central portion of the processes. The vesicle wall is thin, single-layered and ornamented by regular striae that radially spread from the process bases. Additionally, granular ornamentation may occur on the central surface of the vesicle. The surface of the processes is smooth. The processes are hollow inside but it is uncertain whether they are connected through the thickened opaque portion at the base with the vesicle cavity or whether there is a plug. The opening of vesicle (pylome) is circular and large, almost reaching the diameter of the vesicle.

Material. – Five fairly well-preserved specimens.

Dimensions. – $N=5$. Vesicle diameter 36–47 µm (holotype 47 µm), length of processes 7–11 µm (holotype 9–11 µm).

Remarks. – *S. robustum* n.sp. differs from other species of *Stelliferidium* by having more robust and wider processes

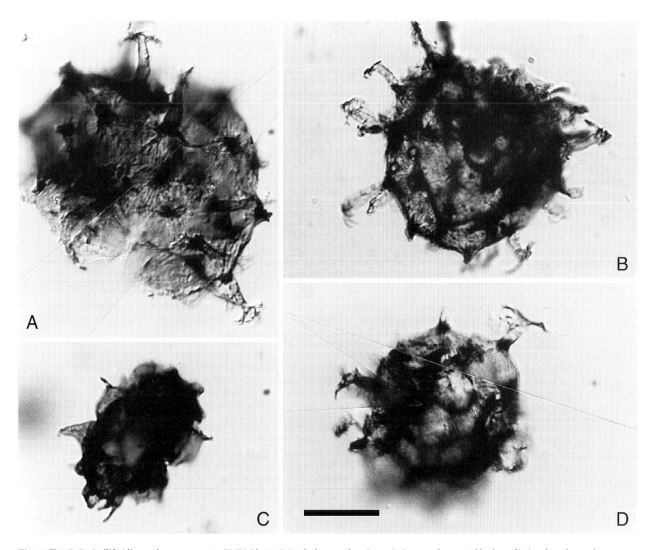

Fig. 45. □A, B, D. *Stelliferidium robustum* n.sp.; A, PMU-Pl.153-R/39, holotype, showing striations on the central body, radiating from base of processes, and robust processes with opaque basal plugs; B, PMU-Pl.154-O/36/4; D, PMU-Pl.155-K/42/3. □C. *Polygonium* sp. A. PMU-Pl.156-O/31/3, differing types of processes, conical and bulbous in shape, freely connected with vesicle cavity. All specimens from the Sosnowiec IG-1 borehole, Sosnowiec formation. A, B, D, depth 3166.3–3167.7 m, and C, depth 3174.0–3174.7 m, lower Upper Cambrian. Scale bar in D equals 14 µm for A and B; 12 µm for C and D.

possessing conspicuous conical bases and thickenings of the process wall between the bases and the main portions of processes. The wall thickenings of the processes become contracted during the process of fossilization, resulting in formation of darker plugs at the basal portion of processes. The processes with more gradual transitions between conical bases and cylindrical portions possess also a dense and darker wall at this transition, suggesting that it is a morphological feature. The apparent polygonal fields observed on the vesicle are caused by the compaction folds and are taphonomic.

Present record. – As for holotype.

Occurrence and stratigraphic record. – Northern Norway, Finnmark, Digermul Penninsula, the Kistedal Formation, Upper Cambrian, *Agnostus pisiformis* Zone to *Acerocare* Zone (Welsch 1986).

Genus *Timofeevia* Vanguestaine, 1978

Type species. – *Timofeevia lancarae* (Cramer & Diez, 1972) Vanguestaine, 1978, p. 272 (=*Multiplicisphaeridium lancarae* Cramer & Diez, 1972, p. 42, Pl. 1:1–4, 6, 8), Spain, the Cantabrian Mountains at Láncara de Luna, upper Middle Cambrian.

Timofeevia lancarae (Cramer & Diez, 1972) Vanguestaine, 1978

Fig. 46A–B

Synonymy. – □invalid 1971 *Archaeohystrichosphaeridium*? sp. A – Gardiner & Vanguestaine, pp. 182–183, Pl. 2:1. □1972 *Multiplicisphaeridium lancarae* Cramer & Diez (New Species) – Cramer & Diez, p. 42, Pl. 1:1–4, 6, 8. □1976 *Baltisphaeridium vilnense* Jankauskas, sp.nov. – Jankauskas, pp. 188–189, Pl. 25:1, 3, 6. □1978 *Timofeevia lancarae* (Cramer et Diez) Vanguestaine nov. comb. – Vanguestaine, p. 272. □1979 *Multiplicisphaeridium vilnense* (Jankauskas, 1976) Jankauskas, comb.nov. – Volkova *et al.*, p. 18, Pl. 1:11, 13. □1991 *Timofeevia lancarae* (Cramer & Diez) Vanguestaine, 1978 – Di Milia, pp. 145–146, Pl. 4:3–4. □1993 *Timofeevia lancarae* (Cramer y Diez) Vanguestaine, 1978 – Fombella *et al.*, Pl. 2:4. □1994 *Timofeevia lancarae* (Cramer & Diez) Vanguestaine, 1978 – Martin *in* Young *et al.*, Fig. 12u. □1995 *Timofeevia lancarae* – Moczydłowska & Crimes, pp. 123–124, Pl. 4A, B.

See additional synonymy lists by Welsch (1986) and Di Milia (1991).

Material. – Ten fairly well-preserved specimens.

Description. – Consistent with the original descriptions of the species by Cramer & Diez (1972, p. 42) and of the genus by Vanguestaine (1978, p. 272).

Dimensions. – *N*=10. Vesicle diameter 23–54 μm, length of processes 7–18 μm.

Present record. – The Upper Silesia area, Sosnowiec IG-1 borehole, Sosnowiec formation at a depth of 3343.9–3347.6 m, Middle Cambrian, *Paradoxides paradoxissimus* Superzone; the Potrójna IG-1 borehole, Jaszczurowa formation at a depth of 3356.3–3363.5 m, Upper Cambrian.

Occurrence and stratigraphic range. – The species has been recorded in Spain, Italy (Sardinia), Czech Republic, Turkey, Morocco, Libya, Tunisia, Ireland, Canada (Newfoundland), Norway, Sweden, Lithuania, Latvia, Estonia and Russia. It ranges from Middle to Upper Cambrian (compilations of occurrences by Welsch 1986 and Di Milia 1991). Additionally, it occurs in south-central Sweden at Kvantorp, 'oelandicus beds', Middle Cambrian *Acadoparadoxides oelandicus* Superzone, *Ptychagnostus praecurrens* Zone (Hagenfeldt 1989b), Wales, in the Maentwrog Formation and Ffestiniog Flags Formation, Upper Cambrian (Martin *in* Young *et al.* 1994), and in Ireland, Co. Wexford, the Booley Bay Formation, upper Upper Cambrian (Moczydłowska & Crimes 1995).

Occurrence of *T. lancarae* in Tremadoc, to which part of the Oville Formation was attributed by Fombella (1978, 1979, 1982, 1986), is not confirmed. The relative age of the Oville Formation in the area concerned is Middle Cambrian (Zamarreño 1972; Aramburu *et al.* 1992; see under 'Stratigraphic ranges').

Occurrence of conspecific microfossils in the presumed Lower Cambrian strata in Lithuania (Jankauskas 1976; Volkova *et al.* 1979) was re-evaluated by Jankauskas (1980). Strata in Lithuania (at Yachenis and the other boreholes, in the Lakaj Formation) and Latvia (Ludza borehole) were subsequently referred to the Middle Cambrian, whereas those in Russia (St. Petersburg area, Izhora River, Izhora Formation) were referred to the Upper Cambrian (Jankauskas 1980).

Timofeevia phosphoritica Vanguestaine, 1978

Fig. 46C–D

Synonymy. – □invalid 1959 *Archaeohystrichosphaeridium Ianischewskyi* sp.n. – Timofeev, p. 33, Pl. 3:2. □invalid 1959 *Archaeohystrichosphaeridium minor* sp.n. – Timofeev, p. 33, Pl. 3:3. □invalid 1971 *Archaeohystrichosphaeridium minor* Timofeev, 1959 – Gardiner & Vanguestaine, p. 182, Pl. 2:3. □invalid 1976 *Cymatiogalea ianishewski* (Timofeev) comb.n. – Vavrdová, p. 60. □invalid 1976 *Cymatiogalea minor* (Timofeev) comb.n. – Vavrdová, p. 60. □1978 *Timofeevia phosphoritica* Vanguestaine, 1978, n.sp. – Vanguestaine, pp. 272–274, Pl. 3:1–8, 10–12. □1990 *Timofeevia janischewskyi* (Timofeev, 1959) Volkova, comb.nov. – Volkova, pp. 84–85, Pls. 4:1–5; 5:1, 3. □1991 *Timofeevia phosphoritica* Vanguestaine, 1978 – Di Milia, p. 147, Pl. 4:7–17. □1991 *Timofeevia phosphoritica* Vanguestaine, 1978 – Albani *et al.*, p. 277, Pl. 3:6–8, 10. □1993 *Timofeevia phosphoritica* Vanguestaine, 1978 – Fombella *et al.*, Pl. 3:5. □1993 *Timofeevia phosphoritica* Vanguestaine, 1978 – Rebecai & Vanguestaine, p. 55, Pl. 2:8. □1994 *Timofeevia phosphoritica* Vanguestaine, 1978 – Martin *in* Young *et al.*, Fig. 12 q. □1995 *Timofeevia phosphoritica* Vanguestaine, 1978 – Moczydłowska & Crimes, Fig. 8:3–4.

See additional synonymy by Albani *et al.* (1991, p. 277).

Material. – One-hundred-and-seven relatively well-preserved specimens.

Fig. 46. □A–B. *Timofeevia lancarae* (Cramer & Diez) Vanguestaine; A, PMU-Pl.157-U/39; B, PMU-Pl.158-R/33. □C–D. *Timofeevia phosphoritica* Vanguestaine. C, PMU-Pl.159-I/44/4; D, PMU-Pl.160-R/33/1. □E–F. *Timofeevia pentagonalis* (Vanguestaine) Vanguestaine. E, PMU-Pl.161-K/45/4; F, PMU-Pl.162-W/49/3. Specimens in A and F from the Potrójna IG-1 borehole, depth 3356.3–3363.5 m, Jaszczurowa formation, Upper Cambrian; all others from the Sosnowiec IG-1 borehole, Sosnowiec formation. B, C, D, depth 3343.9–3347.6 m, Middle Cambrian, *Paradoxides paradoxissimus* Superzone, E, depth 3166.3–3167.7 m, lower Upper Cambrian. Scale bar in E equals 15 μm for A and B; 12 μm for C–F.

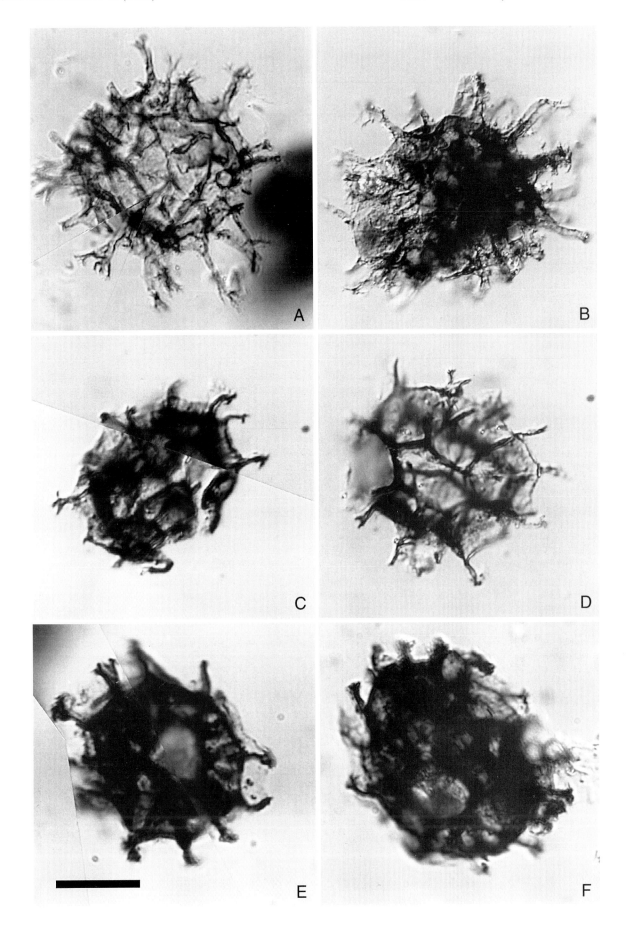

Description. – According to the original diagnosis by Vanguestaine (1978, pp. 272–274).

Dimensions. – N=20. Vesicle diameter 19–34 μm, length of processes 4–6 μm.

Remarks. – The taxonomic re-evaluation of the species by Volkova (1990) is not followed here, because the proposed new combination, as *Timofeevia janischewskyi* (Timofeev, 1959) Volk. comb.nov., was based on a taxon invalidly published by Timofeev (1959) (Moczydłowska & Crimes 1995).

Present record. – The Upper Silesia area, Sosnowiec IG-1 borehole, Sosnowiec formation at depths of 3353.3–3359.3 m and 3343.9–3347.6 m, Middle Cambrian, *Acadoparadoxides oelandicus* and *Paradoxides paradoxissimus* Superzones, respectively; Potrójna IG-1 borehole, Jaszczurowa formation at a depth of 3356.3–3363.5 m, Upper Cambrian.

Occurrence and stratigraphic range. – The species has been recorded in Middle and Upper Cambrian strata in Belgium, France, Spain, Czech Republic, Ireland, England, Sweden, Norway, Italy (Sardinia), Algeria, Libya, Turkey and Canada (Newfoundland) (see compilation by Di Milia 1991). Volkova (1990) and Paalits (1992) reported the species in Russia and Estonia in the Middle–Upper Cambrian and Lower Tremadoc. Additional occurrences are in Belgium and France in the Revinien Group, Middle and Upper Cambrian (Ribecai & Vanguestaine (1993), and in Wales, the Maentwrog Formation and Ffestiniog Flags Formation, Upper Cambrian (Martin in Young *et al.* 1994). Moczydłowska & Crimes (1995) provided the new record of the species in Ireland, Co. Wexford, the Booley Bay Formation, upper part of the Upper Cambrian.

Timofeevia pentagonalis (Vanguestaine, 1974) Vanguestaine, 1978

Fig. 46E–F

Synonymy. – □1974 *Polyedrixium? pentagonale* n.sp. – Vanguestaine, pp. 75–76, Pl. 2:1. □1978 *Timofeevia pentagonalis* (Vanguestaine, 1974) nov.comb. – Vanguestaine, p. 272, Pl. 3:17–21, 23–25. □1981 *Timofeevia pentagonalis* (Vanguestaine, 1974) Vanguestaine, 1978 – Martin & Dean, p. 21, Pl. 5:7 and 9. □1986 *Timofeevia pentagonalis* (Vanguestaine 1974) – Welsch, pp. 86–87, Pl. 4:7–8. □1988 *Timofeevia pentagonalis* (Vanguestaine) Vanguestaine, 1978 – Martin & Dean, Pl. 12:5–6. □1988 *Timofeevia pentagonalis* (Vanguestaine) Vanguestaine, 1978 – Bagnoli *et al.*, p. 205, Pl. 29:2–3. □1991 *Timofeevia pentagonalis* (Vanguestaine) Vanguestaine, 1978 – Di Milia, p. 146, Pl. 4:5–8. □1991 *Timofeevia pentagonalis*

(Vanguestaine) Vanguestaine, 1978 – Albani *et al.*, pp. 276–277, Pl. 3:9. □1993 *Timofeevia pentagonalis* (Vanguestaine) Vanguestaine – Ribecai & Vanguestaine, Pl. 2:9.

Material. – Three specimens satisfactorily preserved.

Description. – Consistent with the description of the species and comparisons by Vanguestaine 1974 (pp. 75–76) and Vanguestaine 1978 (pp. 272–274).

Dimensions. – N=3. Vesicle diameter 23–34 μm, length of processes 4–5 μm, diameter of the polygonal fields 8–9 μm.

Present record. – The Upper Silesia area, Sosnowiec IG-1 borehole, Sosnowiec formation at a depth of 3166.3–3167.7 m, and Potrójna IG-1 borehole, Jaszczurowa formation at a depth of 3356.3–3363.5 m, Upper Cambrian.

Occurrence and stratigraphic range. – The species has been recorded in Belgium, France, Italy (Sardinia), Norway, Sweden, Canada (Newfoundland) and Libya in the uppermost Middle Cambrian, *Paradoxides forchhammeri* Superzone, and Upper Cambrian, *Agnostus pisiformis* Zone to *Acerocare* Zone (see compilations by Welsch 1986 and Di Milia 1991).

Genus *Vogtlandia* Burmann, 1970

Type species. – *Vogtlandia ramificata* Burmann, 1970, pp. 292–293, Pl. 3:4, 5; Germany, Early Ordovician.

Vogtlandia simplex n.sp.

Fig. 36D

Holotype. – Specimen PMU-Pl.114-L/37; Fig. 36D.

Derivation of name. – From Latin *simplex* – simple, single. It refers to the morphology of the processes, dichotomizing in a simple pattern, i.e. from the same point.

Locus typicus. – The Upper Silesia area, Sosnowiec IG-1 borehole (Figs. 2 and 3).

Stratum typicum. – Alternating mudstones and sandstones of the Sosnowiec formation at a depth of 3174.0–3174.7 m; lower Upper Cambrian.

Material. – Two well-preserved specimens.

Diagnosis. – Vesicle originally spherical to polyhedral, bearing not very numerous (12–14 were observed on the vesicle contour), long processes with terminations ramifying from the same point. The processes are tubular and tapering towards the distal portions. Their terminations are divided into three or four branches almost perpendicular to the processes. Single or few branches may have pinnae of the second order on their tips. The processes are hollow and freely connected with the vesicle cavity.

Dimensions. – N=2. Diameter of vesicle 9–18 μm, length of processes 9–13 μm.

Remarks. – The genus *Vogtlandia* was regarded a junior synonym of *Multiplicisphaeridium* by Eisenack *et al.* 1976, but it was retained as a separate genus by Martin (*in* Dean & Martin 1978 and Martin 1983) (see Fensome *et al.* 1990). Volkova (1990) recognized a new species of *Vogtlandia*, *V. notabilis* Volkova, 1990, and proposed a new combination for another species, *V. petropolitana* (German, 1974) Volkova, 1990 (German *in* German & Timofeev 1974). Subsequently, she considered (Volkova 1993, p. 17) *V. notabilis* to be a junior synonym of '*Vogtlandia cervinacornua* (Welsch)'. The latter informal combination was referred to *Multiplicisphaeridium cervinacornua* Welsch, 1986 (Welsch 1986, pp. 61–62, Pl. 6:7–10) which name was corrected as *M. cervinacornuum* Welsch, 1986 (Fensome *et al.* 1990, p. 342). However, the new combination '*V. cervinacornua* (Welsch)' has not been formally proposed, though this illegitimate name was used (Volkova 1993, list of species and pp. 17 and 22). Undoubtedly, the two species, *M. cervinacornuum* Welsch, 1986, and *V. notabilis* Volkova, 1990, are synonymous and the combination *Vogtlandia cervinacornua* (Welsch, 1986) Volkova, 1993 is validated here. The holotype is indicated by Welsch (1986, p. 61, Pl. 6:8). The range of the species is the *Obolus* Zone, being included in the Lower Tremadoc (Welsch 1986) or Upper Cambrian (Volkova 1993).

The genus *Vogtlandia* differs from *Multiplicisphaeridium* by having fewer processes, which are longer and larger in relation to the vesicle dimensions. The length of the processes is usually equal to or greater than the vesicle diameter. The general habit of *Multiplicisphaeridium* is that the vesicle is distinct from the processes, being a dominant part of the specimens, whereas in *Vogtlandia* the processes are more prominent and their bases coalesce to form the vesicle.

The new species described here differs from *Vogtlandia flos* Martin, 1978 (in Dean & Martin 1978), by having branches of the processes almost in one plane, contrary to those in *V. flos*, which are bent in shape of an anchor. It has also smaller dimensions than previously recognized species.

Present record. – As for the holotype.

Vogtlandia sp.
Fig. 36C

Material. – A single well-preserved specimen.

Description. – Vesicle polygonal in outline, bearing a few large processes extending from the wide conical bases that form an angular outline of the vesicle. The processes are equal to or longer than the vesicle diameter. They are cylindrical and show multiple dichotomy, to the 3rd–5th order, in the distal portions of the processes. The processes are hollow and freely connected with the vesicle cavity.

Dimensions. – Diameter of vesicle 14 μm, length of processes 15 μm, width of processes around 2 μm.

Remarks. – The conical bases of some processes appear to have a thickened wall or a dark, thick plug. This thickening is irregular in shape and seems to be a taphonomic feature.

Present record. – The Upper Silesia area, Sosnowiec IG-1 borehole, Sosnowiec formation at a depth of 3166.3–3167.7 m, Upper Cambrian.

Conclusions

The present study has revealed the abundant occurrence of acritarchs in the sedimentary successions sampled in the Sosnowiec IG-1, Goczałkowice IG-1 and Potrójna IG-1 boreholes in Upper Silesia. The results can be summarized as follows:

- The record of microfossils indicates the presence of Lower, Middle and Upper Cambrian strata in a normal stratigraphic succession underlying Devonian strata and overlying Proterozoic basement complexes.

- Observations on sedimentary structures, facies relationships, taphonomy of microfossils and thermal maturation of organic matter indicate that the microfossil assemblages form *in situ* accumulations in the depositional settings concerned.

- The state of preservation of microfossils is generally good, including three-dimensional preservation, and the thermal maturation of organic matter is low to moderate. Values of the thermal alteration index of organic matter suggest proto- and mesocatagenesis stages of lithogenesis.

- The sedimentary succession in the Sosnowiec IG-1 borehole (the informal Sosnowiec formation) is referred to the Middle and Upper Cambrian. The *Acadoparadoxides oelandicus* and *Paradoxides paradoxissimus* Superzones, and probably the *Paradoxides forchhammeri* Superzone, are recognized within the Middle Cambrian strata on the basis of acritarchs. The Upper Cambrian is undetermined, but the sequence probably represents its lower part .

- In the Goczałkowice IG-1 succession (the informal Goczałkowice formation), the Lower Cambrian, previously recognized in part of the succession containing *Holmia*-age diagnostic trilobites, is additionally extended into the *Schmidtiellus mickwitzi* and *Holmia kjerulfi* Zones, and probably the *Protolenus* Zone. The Middle Cambrian, time equivalent to the *Acadoparadoxides oelandicus* Superzone, is for the first time recognized in the upper portion of the succession. The Lower–Middle Cambrian boundary occurs within the sedimentologically continuous strata, thus suggesting that the Upper Silesia terrane was not affected by the Hawke Bay regression.

- Red beds at the base of the Goczałkowice succession (the Pszczyna member) are similar and probably coeval to a sequence at the base of the Potrójna succession (the Potrójna Formation) that was previously interpreted as a Late Proterozoic molasse. Both are inferred to be Vendian in age and represent post-Cadomian molasse deposits in the Upper Silesia terrane.

- The upper portion of the Potrójna IG-1 succession (the informal Jaszczurowa formation), the only one containing acritarchs, is considered as Upper Cambrian.

- Resulting from the present taxonomic revision of certain acritarchs and the re-evaluation of stratigraphic ranges the relative age of the Oville Formation in the Cantabrian Mountains, northern Spain, which yielded acritarchs, is suggested to be Middle Cambrian. The UmbriaPipeta Formation in Sierra Morena, western Spain, is here considered to be Late Cambrian in age.

- The co-occurrence of diverse morphotypes and variable dimensional clusters of microfossils in different depositional settings in Upper Silesia do not support the existence of biofacies restricted to particular depositional environments within a shallow marine shelf.

- Most of the recorded species are clearly cosmopolitan and common to Baltica, Avalonia, Armorica, and Laurentia. This provides additional evidence suggesting that there were no distinctive acritarch bioprovinces during Cambrian times. The worldwide distribution of numerous planktic taxa indicates the absence of palaeoenvironmental barriers, a feature that resulted into free plankton dispersal between the contiguous shelves of various palaeocontinents.

- Biodiversity trends recorded in Upper Silesia reflect global fluctuations in phytoplankton diversity during Cambrian times. Some discrepancies in the diversity pattern, observed at the Lower–Middle Cambrian transition, are interpreted to depend on regional features.

- Major bio-events recorded in the global phytoplankton diversity during Cambrian are:

- Major radiations took place at the beginning of Cambrian, i.e. in the *Platysolenites* Biochron, in the Early Cambrian *Holmia kjerulfi* Biochron, and in the Late Cambrian *Agnostus pisiformis* Biochron.

- There were major turnovers of phytoplankton in the Middle Cambrian *Acadoparadoxides oelandicus* Biochron, and in the Late Cambrian *Peltura* Biochron.

- There was a two-staged extinction at the Early–Middle Cambrian transition, i.e. in the *Protolenus* and *A. oelandicus* Biochrons, and one less significant extinction in the Middle Cambrian *Paradoxides paradoxissimus* Biochron.

- The record of contemporaneous Cambrian geo- and bio-events occurring in short time intervals of few million years provides an indication that many of the observed evolutionary changes were environmentally triggered.

- The duration of the two lowermost zones in the Lower Cambrian, the *Asteridium tornatum – Comasphaeridium velvetum* and *Skiagia ornata – Fimbriaglomerella membranacea* acritarch Zones, corresponding to the *Platysolenites antiquissimus* and *Schmidtiellus mickwitzi* faunal Zones, respectively, is estimated to 2–5 Ma per zone. This is inferred from the isotopic age determinations of rock units at the Precambrian–Cambrian boundary and in the Lower Cambrian, and from the acritarch-based correlation of the Lower Cambrian strata.

- The red beds underlying the Cambrian strata and overlying the metamorphosed basement and previously recognized as molasse deposits are interpreted to be post-Cadomian molasse.

- Based on the recognition of the Cadomian basement in Upper Silesia and similarities of the acritarch and trilobite associations to the East Avalonia Terranes, it is inferred that the Upper Silesia terrane was located within East Avalonia facing Armorica during Cambrian times.

- The homogenous pattern of acritarch distribution in shelf areas in Baltica, Laurentia, Avalonian and

Armorican margin of Gondwana, and Anabar shelf of Siberia in Early Cambrian times indicates absence of palaeoenvironmental barriers allowing free dispersal of phytoplankton along the shelf areas concerned.

- Pericratonic basins were extending along contiguous shelves facing the Iapetus Ocean and the Avalonian seaway. The latter extended between the Avalonian and Armorican margins of Gondwana, the Finnmarkian shelf of Baltica and the Anabar shelf of Siberia.

Acknowledgements. – I acknowledge founding by the Swedish National Science Research Council (NFR) through grants to M. Moczydłowska and the late G. Vidal, and an NFR post-doctoral research fellowship at Harvard University. The State Geological Institute in Warsaw and Kielce are thanked for access to drillcore material and stratigraphic logs through collaboration within the Europrobe Project, sponsored by the European Science Foundation. Z. Kowalczewski, Kielce, was most helpful in discussing regional geological problems. Z. Szczepanik, Kielce, kindly assisted in the sampling of drillcores in 1992. I am indebted to D. Gee, Uppsala University, for useful discussions on the tectonic development of Phanerozoic Europe. The late Francine Martin (Bruxelles) generously made available to me her collections and shared her knowledge on the taxonomy. I thank A. Knoll, Harvard University, for providing research facilities and for a memorable stay at his lab. Constructive review comments by Stewart Molyneux, British Geological Survey, are greatly acknowledged. The editor, Stefan Bengtson, Swedish Museum of Natural History, is thanked for help and advice throughout the production of this volume. This is a contribution to the Europrobe Project.

In particular, I acknowledge my late husband, Gonzalo Vidal, for stimulating discussions and tireless companionship in search for 'what the hell was going on' in the remote past. This part of my own past I miss intensely.

References

Ahlberg, P. & Bergström, J. 1993: The trilobite *Calodiscus lobatus* from the Lower Cambrian of Scania, Sweden. *Geologiska Föreningens i Stockholm Förhandlingar 115*, 331–334.

Albani, R. 1989: Ordovician (Arenigian) Acritarchs from the Solanas Sandstone Formation, Central Sardinia, Italy. *Bollettino della Società Paleontologica Italiana 28*, 3–37.

Albani, R., Massa, D. & Tongiorgi, M. 1991: Palynostratigraphy (acritarchs) of some Cambrian beds from the Rhadames (Ghadamis) Basin (Western Libya – Southern Tunisia). *Bollettino della Società Paleontologica Italiana 30*, 255–280.

Aleksandrowski, P. 1994: Discussion on 'U–Pb ages from SW Poland: evidence for a Caledonian suture zone between Baltica and Gondwana'. *Journal of Geological Society, London 151*, 1049–1050.

Amard, B. 1992: Ultrastructure of *Chuaria* (Walcott) Vidal and Ford (Acritarcha) from the Late Proterozoic Pendjari Formation, Benin and Burkina-Faso, West Africa. *Precambrian Research 57*, 121–123.

Aramburu, C., Truyols, J., Arbizu, M., Méndez-Bedia, I., Zamarreño, I., Garcia-Ramos, J.C., Suarez de Centi, C. & Vanezuela, M. 1992: El Paleozoico Inferior de la Zona Cantábrica. *In* Gutiérrez Marco, J.G., Saavedra, J. & Rábano, I. (eds.): *Paleozoico Inferior de Ibero-América*, 387–421. Imprime Graficas Topacio, S.A., Madrid.

Astini, R., Hatcher, T. & Hatcher, R. 1995 [reported by Kerr, R.A.]: Missing chunk of North America found in Argentina. *Science 270*, 1567–1568.

Bagnoli, G., Stouge, S. & Tongiorgi, M. 1988: Acritarchs and conodonts from the Cambrian–Ordovician Furuhäll (Köpingsklint) section (Öland, Sweden). *Rivista Italiana di Paleontologia e Stratigrafia 94*, 163–248.

Baldis, B. & Bordonaro, O. 1985: Variaciones de facies en la cuenca Cambrica de la Precordillera Argentina, y su relación con la génesis del borde continental. *Sexto Congreso Latino-Americano de Geología, Bogota Colombia 1*, 149–161.

Baudet, D., Aitken, J.D. & Vanguestaine, M. 1989: Palynology of uppermost Proterozoic and lowermost Cambrian formations, central Mackenzie Mountains, northwestern Canada. *Canadian Journal of Earth Sciences 26*, 129–148.

Bengtson, S. & Conway Morris, S. 1992: Early radiation of biomineralizing phyla. *In* Lipps, J.H. & Signor, P.W. (eds.): *Origin and Early Evolution of the Metazoa*, 447–481. Plenum, New York, N.Y.

Bergström, J. 1989: The origin of animal phyla and the new phylum Procoelomata. *Lethaia 22*, 259–269.

Bergström, J. 1994: Ideas on early animal evolution. *In* Bengtson, S. (ed.): *Early Life on Earth. Nobel Symposium No. 84*, 460–466. Columbia University Press, New York, N.Y.

Bergström, J. & Gee, D.G. 1985: The Cambrian in Scandinavia. *In* Gee, D.G. & Sturt, B.A. (eds.): *The Caledonide Orogen – Scandinavia and Related Areas, Part 1*, 247–271. Wiley, Chichester.

Berthelsen, A. 1992a: Mobile Europe. *In* Blundell, D., Freeman, R. & Mueller, S. (eds.): *A Continent Revealed The European Geotraverse*, 11–32. Cambridge University Press, Cambridge.

Berthelsen, A. 1992b: From Precambrian to Variscan Europe. *In* Blundell, D., Freeman, R. & Mueller, S. (eds.): *A Continent Revealed The European Geotraverse*, 153–164. Cambridge University Press, Cambridge.

Berthelsen, A. 1993: Where different geological philosophies meet: The Trans-European Suture Zone. *In* Gee, D.G. & Beckholmen, M. (eds.): Europrobe Symposium Jablonna 1991. *Publications of the Institute of Geophysics Polish Academy of Sciences A-20:255*, 19–31.

Bielewicz, H., Bielewicz, R. & Ślączka, A. 1985: Proterozoik. *In* Ślączka, A. (ed.): *Profile głębokich otworów wiertniczych Instytutu Geologicznego, Potrójna IG 1, Zeszyt 59*, 41–43. Wydawnictwa Geologiczne, Warszawa.

Biernat, G. & Baliński, A. 1973: Fauna z otworów wiertniczych Sosnowiec IG-1 i Goczałkowice IG-1 (Stromatopiroidea, Tabulata, Brachiopoda i Trilobita). *Kwartalnik Geologiczny 17*, 629–630.

[Biernat, G., Osmólska, H., Kaźmierczak, J. & Baliński, A. 1973: Dokumentacja paleontologiczna osadów dewonu i kambru otworu wiertniczego Goczałkowice IG 1. *In* Kotas, A. (ed.): *Dokumentacja geologiczna wynikowa otworu strukturalno-parametrycznego Goczałkowice IG 1*. Archives Geological Institute, Sosnowiec. Unpublished log of the borehole Goczałkowice IG 1.]

Bordonaro, O. 1992: El Cámbrico de Sudamérica. *In* Gutiérrez Marco, J.G., Saavedra, J. & Rábano, I. (eds.): *Paleozoico Inferior de Ibero-América*, 69–84. Imprime Graficas Topacio, S.A., Madrid.

Bottjer, D.J. & Droser, M.L. 1994: The history of Phanerozoic bioturbation. *In* Donovan, S.K. (ed.): *The Palaeobiology of Trace Fossils*, 155–176. Wiley, Chichester.

Bowring, S.A., Grotzinger, J.P., Isachsen, C.E., Knoll, A.H., Pelechaty, S.M. & Kolosov, P. 1993: Calibrating rates of Early Cambrian Evolution. *Science 261*, 1293–1298.

Brasier, M.D. 1980: The Lower Cambrian transgression and glauconite–phosphate facies in western Europe. *Journal of the Geological Society London 137*, 695–703.

Brasier, M.D. 1989a: China and the Palaeotethyan Belt (India, Pakistan, Iran, Kazakhstan, and Mongolia). *In* Cowie, J.W. & Brasier, M.D (eds.): *The Precambrian–Cambrian boundary*, 40–74. Clarendon, Oxford.

Brasier, M.D. 1989b: Sections in England and their correlation. *In* Cowie, J.W. & Brasier, M.D (eds.): *The Precambrian–Cambrian boundary*, 82–104. Oxford Science Publications, Oxford.

Brasier, M.D. 1990: Phosphogenic events and skeletal preservation across the Precambrian–Cambrian boundary interval. *In* Notholt, A.J.G. & Jarvis, I. (eds.): Phosphorite research and development. *Geological Society Special Publication 52*, 289–303.

Brasier, M.D. 1991: Nutrient flux and the evolutionary explosion across the Precambrian–Cambrian boundary interval. *Historical Biology 5*, 85–93.

Brasier, M.D. 1992: Paleoceanography and changes in the biological cycling of phosphorus across the Precambrian–Cambrian boundary. *In* Lipps, J.H. & Signor, P.W. (eds.): *Origin and Early Evolution of the Metazoa*, 483–523. Plenum, New York, N.Y.

Brochwicz-Lewiński, W., Pożaryski, W. & Tomczyk, H. 1981: Wielkoskalowe ruchy przesuwcze wzdłuż SW brzegu platformy wschodnioeuropejskiej we wczesnym paleozoiku. *Przegląd Geologiczny 29*, 385–397.

Brochwicz-Lewiński, W., Pożaryski, W. & Tomczyk, H. 1983: Ruchy przesuwcze w południowej Polsce w paleozoiku. *Przegląd Geologiczny 31*, 651–658.

Brochwicz-Lewiński, W., Vidal, G., Pożaryski, W., Tomczyk, H. & Zając, R. 1986: Position tectonique du massif de Haute-Silesie avant le Permien a la lumiere de donnees nouvelles sue le Cambrien de cette region. *Comptes Rendus de l'Académie des Sciences Paris 303, Serie II, 16*, 1493–1496.

Bukowy, S. 1982: Problemy budowy paleozoiku regionu śląsko-krakowskiego. *In* Różkowski, A. & Ślósarz, J. (eds.): *Przewodnik LIV Zjazdu Polskiego Towarzystwa Geologicznego, Sosnowiec 23–25 września 1982*, 7–26. Wydawnictwa Geologiczne, Warszawa.

Bukowy, S. 1994: Zarys budowy paleozoiku północno-wschodniego obrzeżenia Górnośląskiego Zagłębia Węglowego. *In* Różkowski, A., Ślósarz, J. & Żaba, J. (eds.): *Przewodnik LXV Zjazdu Polskiego Towarzystwa Geologicznego, Sosnowiec 22–24 września 1994*, 14–30. Wydawnictwo Uniwersytetu Śląskiego, Katowice.

Burmann, G. 1970: Weitere organische Mikrofossilien aus dem unteren Ordovizium. *Paläontologische Abhandlungen, Abt. B, Paläobotanik 3*, 289–332.

Butterfield, N.J., Knoll, A.H. & Swett, K. 1994: Paleobiology of the Neoproterozoic Svanbergfjellet Formation, Spitsbergen. *Fossils and Strata 34*. 84 pp.

Cebulak, S. & Kotas, A. 1982: Profil utworów intruzywnych i prekambryjskich w otworze Goczałkowice IG 1. *In* Różkowski, A. & Ślósarz, J. (eds.): *Przewodnik LIV Zjazdu Polskiego Towarzystwa Geologicznego, Sosnowiec 23–25 września 1982*, 205–210. Wydawnictwa Geologiczne, Warszawa.

[Cebulak, S., Nurkiewicz, B. & Skupień, M. 1973a: Wyniki badań petrograficznych i mineralogicznych. *In* Kotas, A. (ed.): *Dokumentacja geologiczna wynikowa otworu strukturalno-parametrycznego Goczałkowice IG 1*. Archives State Geological Institute, Warsaw Unpublished log of the borehole Goczałkowice IG-1.]

[Cebulak, S., Nurkiewicz, B. & Skupień, M. 1973b: Wyniki badań petrograficznych i mineralogicznych. *In* Kotas, A. (ed.): *Dokumentacja geologiczna wynikowa otworu strukturalno-parametrycznego Sosnowiec IG 1*. Archives State Geological Institute, Warsaw (Unpublished log of the borehole Sosnowiec IG-1.]

Cebulak, S., Nurkiewicz, B. & Skupień, M. 1973c: Charakterystyka petrograficzna skał w otworze wiertniczym Sosnowiec IG 1 i Goczałkowice IG 1. *Kwartalnik Geologiczny 17*, 372–373.

Cloud, P. 1968a: Atmospheric and hydrospheric evolution on the primitive Earth. *Science 160*, 729–736.

Cloud, P. 1968b: Pre-metazoan evolution and the origins of the Metazoa. *In* Drake, T. (ed.): *Evolution and Environment*, 1–72. Yale University Press, New Haven, Conn.

Cloud, P. 1976: The beginnings of biospheric evolution and their biogeochemical consequences. *Paleobiology 2*, 351–387.

Cloud, P. 1988: *Oasis in Space. Earth history from the Beginning*. 508 pp. Norton, New York, N.Y.

Colbath, G.K. 1979: Organic-walled microphytoplankton from the Eden Shale (Upper Ordovician), Indiana, U.S.A. *Palaeontographica B 171*, 1–38.

Colbath, G.K. 1983: Fossil prasinophycean phycomata (Chlorophyta) from the Silurian Bainbridge Formation, Missouri, U.S.A. *Phycologia 22*, 249–265.

Colbath, G.K. 1990: Devonian (Givetian–Frasnian) organic-walled phytoplankton from the Limestone Billy Hills reef complex, Canning Basin, Western Australia. *Palaeontographica B 217*, 87–145.

Colbath, G.K. & Grenfell, H.R. 1995: Review of biological affinities of Paleozoic acid-resistant, organic-walled eukaryotic algal microfossils (including 'acritarchs'). *Review of Palaeobotany and Palynology 86*, 287–314.

Compston, W., Sambridge, M.S., Reinfrank, R.F., Moczydłowska, M., Vidal, G. & Claesson, S. 1995: Numerical ages of volcanic rocks and the earliest faunal zone within the Late Precambrian of east Poland. *Journal of the Geological Society, London 152*, 599–611. .

Conway Morris, S. 1987a: Cambrian enigmas. *Geology Today May–June 1987*, 88–92.

Conway Morris, S. 1989: Burgess Shale Faunas and the Cambrian explosion. *Science 246*, 339–346.

Conway Morris, S. 1989: Southeastern Newfoundland and adjacent areas (Avalon Zone). *In* Cowie, J.W. & Brasier, M.D (eds.): *The Precambrian–Cambrian boundary*, 5–39. Oxford Science Publications, Oxford.

Conway Morris, S. 1992: Burgess Shale-type faunas in the context of the 'Cambrian explosion': A review. *Journal of the Geological Society, London 149*, 631–636.

Conway Morris, S. 1994: Early metazoan evolution: First steps to an integration of molecular and morphological data. *In* Bengtson, S. (ed.): *Early Life on Earth. Nobel Symposium 84*, 450–459. Columbia University Press, New York, N.Y.

Cookson, I.C. & Eisenack, A. 1958: Microplankton from Australian and New Guinea Upper Mesozoic sediments. *Proceedings of the Royal Society of Victoria 70*, 19–79.

Courjault-Radé, P., Debrenne, F. & Gandin, A. 1992: Palaeogeographic and geodynamic evolution of the Gondwana continental margins during the Cambrian. *Terra Nova 4*, 657–667.

Courties, C., Vaquer, A., Troussellier, M, Lautier J., Chretiennot-Dinet, M.J., Neveux, J., Machado, C. & Claustre H. 1994: Smallest eukaryotic organism. *Nature 370*, 255.

Coward, M.P. & Potts, G.J. 1985: Fold nappes: examples from the Moine Thrust zone. *In* Gee, D.G. & Sturt, B.A. (eds.): *The Caledonide Orogen – Scandinavia and Related Areas 2*, 1147–1158. Wiley, Chichester.

Cramer, F.H. 1970: Distribution of selected Silurian acritarchs. An account of the palynostratigraphy and paleogeography of selected Silurian acritarch taxa. *Revista Española de Micropaleontología, Numero Extraordinario*, 1-203.

Cramer, F.H. & Diez Cramer, M.C.R. 1972: Acritarchs from the upper Middle Cambrian Oville Formation of León, northwestern Spain. *Revista Española de Micropaleontología 30*, 39–50.

Cramer, F.H. & Diez, M.d.C.R. 1979: Lower Paleozoic acritarchs. *Palinologia 1*, 17–160.

Crimes, T.P. & Crossley, J.D. 1968: The stratigraphy, sedimentology, ichnology and structure of the Lower Palaeozoic rocks of part of northeastern Co. Wexford. *Proceedings of the Royal Irish Academy 67B*, 185–215.

Crimes, T.P., Insole, A. & Williams, B.P.J. 1995: A rigid-bodied Ediacaran biota from Upper Cambrian strata in Co. Wexford, Eire. *Geological Journal 30*, 89–109.

Dadlez, R., Kowalczewski, Z. & Znosko, J. 1994: Some key problems of the pre-Permian tectonics of Poland. *Geological Quarterly 38*, 169–190.

Dadlez, R. 1995: Debates about the pre-Variscan tectonics of Poland. *Studia Geophysica et Geodaetica 39*, 227–234.

Dallmeyer, R.D. & Martinez Garcia, E. 1990: *Pre-Mesozoic Geology of Iberia*. 416 pp. Springer, Berlin.

Dalziel, I. 1995: Palaeogeographic map of Gondwana in Early Cambrian. *In* Palmer, A.R & Rowell, A.J.: Early Cambrian trilobites from the Ahacleton Limestone of the Central Transatlantic Mountains. *Journal of Paleontology 69, Supp. 6, The Paleontological Society Memoir 45*, p. 5.

Dean, W.T. 1985: Relationships of Cambrian–Ordovician faunas in the Caledonide–Appalachian region, with particular reference to trilobites. *In* Gayer, R.A. (ed.): *The Tectonic Evolution of the Caledonide–Appalachian Orogen*, 17–47. Vieweg, Braunschweig.

Dean, W.T. & Martin, F. 1978: Lower Ordovician acritarchs and trilobites from Bell Island, eastern Newfoundland. *Geological Survey of Canada Bulletin 284*, 1–34.

Dean, W.T. & Martin, F. 1982: The sequence of trilobite faunas and acritarch microfloras at the Cambrian–Ordovician boundary, Wilcox Pass, Alberta, Canada. *In* Bassett, M.G. & Dean, W.T. (eds.): The Cambrian–Ordovician boundary: sections, fossil distributions, and correlations. *National Museum of Wales, Geological Series 3*, 131–140. Cardiff.

Deflandre, G. 1937: Microfossiles des silex Crétacés. Deuxième partie. Flagélles incertae sedis Hystrichosphaerides sarcodinés. Organisme Divers. *Annales de Paleontologie 26*, 51–103.

Deflandre, G. 1938: Microplancton des mers jurassiques conservé dans les marnes de Villers-sur-Mer (Calvados). Étude liminaire et considérations générales. *Travaux de la Station Zoologique de Wimereux 13*, 147–200.

Deflandre, G. 1954: Systématique des Hystrichosphaerides: sur l'acception du genre *Cymatiosphaera* O.Wetzel. *Compte Rendu Sommaire et Bulletin de la Société Géologique de France 4*, 9–10.

Deflandre, G. 1968: Sur l'existence, dans le Précambrien, d'Acritarches du type Acanthomorphitae: *Eomicrhystridium* nov.gen. Typification du genre *Palaeocryptidium* Defl. 1955. *Comptes Rendus Hebdomadaires des Séances de l'Académie des Sciences. Paris, Ser. D, 266*, 2385–2389.

Deunff, J., Górka, H. & Rauscher, R. 1974: Observations nouvelles et précisions sur les Acritarches a large ouverture polaire du Paleozoique inférieur. *Geobios 7*, 5–18.

Di Milia, A. 1991: Upper Cambrian acritarchs from the Solanas Sandstone Formation, Central Sardinia, Italy. *Bollettino della Societá Paleontologica Italiana 30*, 127–152.

D'Lemos, R.S., Strachan, R.A. & Topley, C.G. (eds.) 1990: *The Cadomian Orogeny.* 423 pp. *The Geological Society Special Publication 51.* London.

Don, J. & Żelaźniewicz, A. 1990: The Sudetes – boundaries, subdivision and tectonic position. *Neues Jahrbuch für Geologie und Paläontologie, Abhandlungen 179*, 121–127.

Downie, C. 1963: 'Hystrichospheres' (acritarchs) and spores of the Wenlock Shales (Silurian) of Wenlock, England. *Palaeontology 6*, 625–652.

Downie, C. 1982: Lower Cambrian acritarchs from Scotland, Norway, Greenland and Canada. *Transactions of the Royal Society of Edinburgh: Earth Sciences 72*, 257–295.

Downie, C. 1984: Acritarchs in British stratigraphy. *Geological Society, London, Special Report 17.* 26 pp.

Downie, C., Evitt, W.R. & Sarjeant, W.A.S. 1963: Dinoflagellates, hystrichospheres, and the classification of the acritarchs. *Stanford University Publications in Geological Sciences 7*, 1–16.

Droser, M.L. 1991: Ichnofabrics of the Paleozoic *Skolithos* ichnofacies and the nature and distribution of piperock. *Palaios 6*, 316–325.

Eisenack, A. 1958a: *Tasmanites* Newton 1875 und *Leiosphaeridia* n.gen. aus Gattungen der Hystrichosphaeridea. *Palaeontographica Abt. A 110*, 1–19.

Eisenack, A. 1958b: Mikroplankton aus dem norddeutschen Apt. *Neues Jahrbuch für Geologie und Paläontologie, Abhandlungen 106*, 383–422.

Eisenack, A. 1969: Zur Systematic einiger paläozoischer Hystrichosphären (Acritarcha) des baltischen Gebietes. *Neues Jahrbuch für Geologie und Paläontologie, Abhandlungen 133*, 245–266.

Eisenack, A. 1972: Kritische Bemerkung zur Gattung *Pterospermopsis* (Chlorophyta, Prasinophyceae). *Neues Jahrbuch für Geologie und Paläontologie, Monatshefte 10*, 596–601.

Eisenack, A., Cramer, F.H. & Diez Rodrigez, M.d.C. 1973: *Katalog der fossilen Dinoflagellaten, Hystrichosphären und verwandten Mikrofossilien. Band III, Acritarcha, Teil 1.* 1104 pp. Stuttgart.

Eisenack, A., Cramer, F.H. & Diez, M.C.R. 1976: *Katalog der fossilen Dinoflagellaten, Hystrichosphären und verwandten Mikrofossilien. Band IV, Acritarcha, Teil 2.* 863 pp. Stuttgart.

Eisenack, A., Cramer, F.H. & Diez, M.C.R. 1979a: *Katalog der fossilen Dinoflagellaten, Hystrichosphären und verwandten Mikrofossilien. Band V Acitarcha, Teil 3.* 532 pp. Stuttgart.

Eisenack, A., Cramer, F.H. & Diez, M.C.R. 1979b: *Katalog der fossilen Dinoflagellaten, Hystrichosphären und verwandten Mikrofossilien. Band VI Acitarcha, Teil 3.* 533 pp. Stuttgart.

Eklund, K. 1990: Lower Cambrian acritarch stratigraphy of the Bårstad 2 core, Östergötland, Sweden. *Geologiska Föreningens i Stockholm Förhandlingar 112*, 19–44.

Erdtmann, B.-D.1991: The post-Cadomian Early Palaeozoic tectonostratigraphy of Germany (Attempt at an analytical review). *Annales Société Géologique Belgique 114*, 19–43.

Erkmen, U. & Bozdoğan, N. 1981: Cambrian acritarchs from the Sosink Formation in southeast Turkey. *Revista Española de Micropaleontologia 13*, 47–60.

Evitt, W.R. 1963: A discussion and proposals concerning fossil Dinoflagellates, Hystrichospheres and Acritarchs. *(U.S) National Academy of Sciences Proceedings 49*, 158–164, 298–302.

Fatka, O. 1989: Acritarch assemblage in the Onymagnostus hybridus Zone (Jince Formation, Middle Cambrian, Czechoslovakia). *Věstnik Ústředniho ústavu geologického 64*, 363–368.

Fensome, R.A., Williams, G.L., Barss, M.S., Freeman, J.M. & Hill, J.M. 1990: Acritarchs and fossil Prasinophytes: and index to genera, species and infraspecific taxa. *American Association of Stratigraphic Palynologists Contributions Series Number 25.* 771 pp.

Fombella, M.A. 1977: Acritarcos de edad Cambrico Medio-inferior de la provincia de León, Española. *Revista Española de Micropaleontologia 9*, 115–124.

Fombella, M.A. 1978: Acritarcos de la Formación Oville, edad Cámbrico Medio – Tremadoc, Provincia de León, España. *Palinologia Numero Extraordinario 1*, 245–261.

Fombella, M.A. 1979: Palinologia de la Formación Oville al Norte y Sur de la Cordillera Cantabrica, España. *Palinologia 1*, 1–14.

Fombella, M.A. 1982: Determinacion palinologica del Tremadoc en la localidad de Verdiago, Provincia de León, NO de España. *Revista Española de Micropaleontologia 14*, 13–22.

Fombella, M.A. 1986: El transito Cambrico–Ordovicico, palinologia y diacronismo, Provincia de León, NO de España. *Revista Española de Micropaleontologia 18*, 165–179.

Fombella, M.A. 1987: Resemblances and differences between the palynological associations of Upper Cambrian age in the NW of Spain (Vozmediano) and north of Africa. *Revue de Micropaléontologie 30*, 111–116.

Fombella Blanco, M.A., Valencia Barrera, R.M., Fernández González, D. & Cachán Santos, L.J. 1993: Diferencias de composición en las asociaciones de acritarcos de seis localidades de la Formacion Oville (NO de España). Edad Cambrico Medio – Tremadoc Inferior. *Revista Española de Paleontologia 18*, 221–235.

Franke, D. 1995: Caledonian terranes along the southwestern border of the East European Platform – evidence, speculation and open questions. *Studia Geophysica et Geodaetica 39*, 241–256.

Franke, W. 1992: Phanerozoic structures and events in Central Europe. In Blundell, D., Freeman, R. & Mueller, S. (eds.): *A Continent Revealed. The European Geotraverse*, 164–180. Cambridge University Press, Cambridge.

Fridrichsone, A-I. 1971: Akritarkhi *Baltisphaeridium* i gistrikhosfery (?) iz kembrijskikh otlozhenij Latvii. [Acritarchs *Baltisphaeridium* and hystrichosphaers (?) from the Cambrian deposits in Latvia.] *In: Paleontologiya i Stratigrafiya Pribaltiki i Belorusi 3*, 5–22.

Gámez, J.A., Fernández-Nieto, C., Gozalo, R., Liñán, E., Mandado, J. & Palacios, T. 1991: Bioestratigrafia y evolución embiental del Cámbrico de Borobia (Provincia de Soria, Cadena Ibérica Oriental). *Cuaderos do Laboratorio Xelóxico de Laxe 16*, 251–271.

Gardiner, P.R.R. & Vanguestaine, M. 1971: Cambrian and Ordovician microfossils from south-east Ireland and their implications. *Geological Survey of Ireland Bulletin 1*, 163–210.

Gee, D.G. 1972: The regional geological context of the Tåsjö uranium project, Caledonian front, central Sweden. *Sveriges Geologiska Undersökning C 671*, 36 pp.

German, T.N. & Timofeev, B.V. 1974: *Veryhachium* kembriya. (*Veryhachium* from the Cambrian). *In: Mikrofossili proterozoya i rannego paleozoya SSSR*, 13–15, Nauka, Leningrad. (In Russian.)

Geyer, G. 1990: Die marokkanischen Ellipsocephalidae (Trilobita: Redlichiida). *Beringeria 3.* 363 pp.

Geyer, G. & Landing, E. 1995: The Cambrian of the Moroccan Atlas regions. *In* Geyer, G. & Landing, E. (eds.): *Morocco '95, The Lower–Middle Cambrian Standard of Western Gondwana, Beringeria Special Issue 2*, 7–46. Würzburg.

Geyer, G., Landing, E. & Heldmaier, W. 1995: Faunas and depositional environments of the Cambrian of the Moroccan Atlas regions. *In* Geyer, G. & Landing, E. (eds.): *Morocco '95, The Lower–Middle Cambrian Standard of Western Gondwana, Beringeria Special Issue 2*, 47–119. Würzburg.

Gonçálvez, F. & Palacios, T. 1984: Novos elementos Paleontológicos e Estratigráficos sobre o Proterozoico Português da Zona de Ossa-Morena. *Memorias da Academia das Ciéncias de Lisboa. Classe de Ciéncias, 25*, 225–235.

Grotzinger, J.P., Bowring, S.A., Saylor, B.Z. & Kaufman, A.J. 1995: Biostratigraphic and geochronologic constraints on early animal evolution. *Science 270*, 598–604.

Habib, D. 1972: Dinoflagellate stratigraphy, Leg 11, Deep Sea Drilling Project. *In* Hollister, C.D, Ewing, J.I. *et al.*: *Initial Reports of the Deep Sea Drilling Project, Volume 11*, 367–425. U.S. Government Printing Office, Washington, D.C.

Habib, D. 1979: Sedimentology of palynodebris in Cretaceous carbonaceous facies south of Vigo Seamount. *In* Sibuet, J.C., Ryan, W.B.F. *et al.*: *Initial Reports of the Deep Sea Drilling Project, Volume 47*, 452–468. U.S. Government Printing Office, Washington, D.C.

Habib, D. & Knapp, S.D. 1982: Stratigraphic utility of Cretaceous small acritarchs. *Micropaleontology 28*, 335–371.

Hagenfeldt, S.E. 1989a: Lower Cambrian acritarchs from the Baltic Depression and south-central Sweden, taxonomy and biostratigraphy. *Stockholm Contributions in Geology 41*, 1–176.

Hagenfeldt, S.E. 1989b: Middle Cambrian acritarchs from the Baltic Depression and south-central Sweden, taxonomy and biostratigraphy. *Stockholm Contributions in Geology 41*, 177–250.

Hagenfeldt, S.E. 1994: The Lower/Middle Cambrian boundary in the Baltic Sea region. *Geologiya 17*, 24–32.

Harańczyk, C. 1982: Nowe dane do poznania kaledońskiego górotworu krakowidów. *In* Różkowski, A. & Ślósarz, J. (eds.): *Przewodnik LIV Zjazdu Polskiego Towarzystwa Geologicznego Sosnowiec 23–25 września 1982*, 90–101. Wydawnictwa Geologiczne, Warszawa, .

Harańczyk, C. 1994: Znaczenie sutury terranowej Zawiercie–Rzeszotary dla poznania kaledońskiego transpresyjnego górotworu krakowidów. *In* Różkowski, A., Ślósarz, J. & Żaba, J. (eds.): *Przewodnik LXV Zjazdu Polskiego Towarzystwa Geologicznego Sosnowiec 22–24 września 1994*, 67–79. Wydawnictwa Uniwersytetu Śląskiego, Katowice.

Hayes, J.M., Kaplan, I.R. & Wedeking, W. 1983: Precambrian Organic Geochemistry, Preservation of the Record. *In* Schopf, J.W. (ed.): *Earth's Earliest Biosphere*, 93–134. Princeton University Press, Princeton, N.J.

Heflik, W. 1982: Utwory metamorficzne z podłoża brzeżnej części Karpat obszaru Cieszyn – Rzeszotary. *In* Różkowski, A. & Ślósarz, J. (eds.): *Przewodnik LIV Zjazdu Polskiego Towarzystwa Geologicznego Sosnowiec 23–25 września 1982*, 210–213. Wydawnictwa Geologiczne Warszawa,

Heflik, W. & Konior, K. 1974: Obecny stan rozpoznania podłoża krystalicznego w obszarze Cieszyn–Rzeszotary. *Biuletyn Instytutu Geologicznego 273*, 196–227.

Hofmann, H.J. 1971: Precambrian fossils, pseudofossils, and problematica in Canada. *Geological Survey of Canada, Bulletin 189*, 1–146.

Hofmann, H.J. 1985a: The mid-Proterozoic Little Dal macrobiota, Mackenzie Mountains, north-west Canada. *Palaeontology 28*, 331–354.

Hofmann, H.J. 1994: Proterozoic carbonaceous compressions ('metaphytes' and 'worms'). *In* Bengtson, S. (ed.): *Early Life on Earth. Nobel Symposium No. 84*, 342–357. Columbia University Press, New York, N.Y.

Ineson, J.R., Surlyk, F., Higgins, A.K. & Peel, J.S. 1994: Slope apron and deep shelf sediments of the Brønlund Fjord and Tavsens Iskappe Groups (Lower Cambrian – Lower Ordovicia), North Greenland: stratigraphy, facies and depositional setting. *Grønlands Geologiske Undersøgelse Bulletin 169*, 7–24.

International Code of Botanical Nomenclature 1988. 1983. 472 pp. Bohn, Scheltema & Holkema, Utrecht.

Isachsen, C.E., Bowring, S.A., Landing, E. & Samson, S.D. 1994: New constraint on the division of Cambrian time. *Geology 22*, 496–498.

Jacobson, S.R. & Achab, A. 1985: Acritarch biostratigraphy of the *Dicellograptus complanatus* graptolite Zone from the Vareal Formation (Ashgillian), Anticosti Island, Quebec, Canada. *Palynology 9*, 165–198.

Jankauskas, T.V. 1974: Korelaciya kembrijskih otlozhenij Litovskoj SSR (po akritarkham). [Correlation of the Cambrian deposits in Lithuanian SSR (acritarch-based).] *In* Zhuravleva, I.T. & Rozanov, A.Yu. (eds.): *Biostratigrafiya i paleontologiya nizhnego kembriya Evropy i severnoj Azii*, 22–29. Nauka, Moscow.

Jankauskas, T.V. 1975: Novye akritarkhi nizhnego kembriya Pribaltiki. [New Lower Cambrian acritarchs from the Peribaltic.] *Paleontologicheskij Zhurnal 1975:1*, 94–104.

Jankauskas, T.V. 1976: Novye vidy akritarkh iz nizhnego kembriya Pribaltiki. [New acritarch species from the Lower Cambrian of the Peribaltic.] *In* Zhuravleva, I.T. (ed.): *Stratigrafiya i paleontologiya nizhnego i srednego kembriya SSSR*, 187–194. Nauka, Novosibirsk.

Jankauskas, T. 1980: K mikrofitologicheskoj kharakteristike srednе- i verkhnekembrijskikh otlozhenij severo-zapadnoj chasti vostochno-evropejskoj platformy. [Microphytological characteristic of the Middle–Upper Cambrian strata in the north-western part of the East European Platform.] *Eesti NSV Teaduste Akadeemia Toimetised, Keemia Geoloogia 29*, 131–135.

Jankauskas, T.V. & Posti, E. 1976: Novye vidy akritarkh kembriya Pribaltiki. [New Cambrian acritarchs from the east Baltic area.] *Eesti NSV Teaduste Akadeemia Toimetised, Keemia Geoloogia 25*, 145–151.

Jensen, S. 1997: Trace fossils from the Lower Cambrian Mickwitzia sandstone, south-central Sweden. *Fossils and Strata 42*, 111 pp.

Johnston, J.D., Tait, J.A., Oliver, G.J.H. & Murphy, F.G. 1994: Evidence for a Caledonian orogeny in Poland. *Transactions of the Royal Society of Edinburgh: Earth Sciences 85*, 131–142.

Johnston, J.D., Tait, J.A., Oliver, G.J.H. & Murphy, F.G. 1996: Reply to Comments by M. Moczydłowska. *Transactions of the Royal Society of Edinburgh: Earth Sciences 86*, 231–232.

Jurkiewicz, H. 1975: Budowa geologiczna podłoża mezozoiku centralnej części Niecki Miechowskiej. *Instytut Geologiczny Biuletyn 283*, 5–99.

Jux, U. 1968: Über den Feinbau der Wandung bei *Tasmanites* Newton. *Palaeontographica B 124*, 112–124.

Jux, U. 1969a: Über den Feinbau der Zystenwandung von *Pachysphaera marshalliae* Parke, 1966. *Palaeontographica B 125*, 104–111.

Jux, U. 1969b: Über den Feinbau der Zystenwandung von *Halosphaera* Schmitz, 1878. *Palaeontographica B 128*, 48– 55.

Jux, U. 1971: Über den Feinbau der Wandungen einiger paläozoischer Baltisphaeridiacean. *Palaeontographica B 136*, 115–128.

Kerr, R.A. 1995: Missing chunk of North America found in Argentina. *Science 270*, 1567–1568.

Kirjanov, V.V. 1974: Novye akritarkhi iz kembrijskikh otlozhenij Volyni. [New acritarchs from the Cambrian deposits of Volhynia.] *Paleontologicheskij Zhurnal 2*, 117–130.

Kjellström, G. 1968: Remarks on the chemistry and ultrastructure of the cell wall of some Palaeozoic leiospheres. *Geologiska Föreningens i Stockholm Förhandlingar 90*, 221–118.

Kjellström, G. 1971: Ordovician Microplankton (Baltisphaerids) from the Grötlingbo Borehole No. 1 in Gotland, Sweden. *Sveriges Geologiska Undersökning C 655*, 1–75.

Knoll, A.H. 1992a: Biological and biogeochemical preludes to the Ediacaran radiation. *In* Lipps, J.H. & Signor, P.W. (eds.): *Origin and Early Evolution of the Metazoa*, 53–84. Plenum Press, New York, N.Y.

Knoll, A.H. 1992b: Life in the Late Proterozoic. *In* Margulis, L. & Olendzenski, L. (eds.): *Environmental Evolution. Effects of the Origin and Evolution of Life on Planet Earth*, 201–213. MIT Press, Cambridge, Mass.

Knoll, A.H. 1994a: Neoproterozoic evolution and environmental change. *In* Bengtson, S. (ed.): *Early Life on Earth. Nobel Symposium No. 84*, 439–449. Columbia University Press, New York, N.Y.

Knoll, A.H. 1994b: Proterozoic and Early Cambrian protists: evidence for accelerating evolutionary tempo. *Proceedings of the National Academy of Sciences U.S.A. 91*, 6743–6750.

Knoll, A.H. & Calder, S. 1983: Microbiotas of the Late Precambrian Ryssö Formation, Nordaustlandet, Svalbard. *Palaeontology 26*, 467–496.

Knoll, A.H. & Holland, H.D. 1995: Oxygen and Proterozoic evolution: an update. *In* Stanley, S.M. (ed.): *Effects of Past Global Change on Life*, 21–33. National Research Council Studies in Geophysics. National Academy Press, Washington, D.C.

Knoll, A.H. & Swett, K. 1987: Micropaleontology across the Precambrian–Cambrian boundary in Spitsbergen. *Journal of Paleontology 61*, 898–926.

Kotas 1972: Osady morskie karbonu górnego i ich przejście w utwory produktywne Górnośląskiego Zagłębia Węglowego. *Prace Instytutu Geologicznego 61*, 279–307.

[Kotas, A. 1973a (ed.): Dokumentacja geologiczna wynikowa otworu strukturalno-parametrycznego Goczałkowice IG 1. Archives State Geological Institute, Warsaw (Unpublished log of the borehole Goczałkowice IG-1).]

[Kotas, A. 1973b (ed.): Dokumentacja geologiczna wynikowa otworu strukturalno-parametrycznego Sosnowiec IG 1. Archives State Geological Institute, Warsaw (Unpublished log of the borehole Sosnowiec IG-1).]

Kotas, A. 1973c: Profil utworów paleozoicznych w otworach wiertniczych Sosnowiec IG-1 i Goczałkowice IG-1. *Kwartalnik Geologiczny 17*, 626–627.

Kotas, A. 1973d: Występowanie kambru w podłożu Górnośląskiego Zagłębia Węglowego. *Przegląd Geologiczny 1*, 57–61.

Kotas, A. 1982a: Zarys budowy geologicznej Górnośląskiego Zagłębia Węglowego. *In* Różkowski, A. &Ślósarz, J. (eds.): *Przewodnik LIV Zjazdu Polskiego Towarzystwa Geologicznego, Sosnowiec 23–25 września 1982*, 45–72. Wydawnictwa Geologiczne, Warszawa.

Kotas, A. 1982b: Profil utworów kambru w otworze Goczałkowice IG 1. *In* Różkowski, A. & Ślósarz, J. (eds.): *Przewodnik LIV Zjazdu Polskiego Towarzystwa Geologicznego, Sosnowiec 23–25 września 1982*, 193–201. Wydawnictwa Geologiczne, Warszawa.

Kowalczewski, Z. 1981: Litostratygrafia wendu w Górach Swiętokrzyskich i Niecce Miechowskiej. *In* Żakowa, H. (ed.): *Przewodnik LIII Zjazdu Polskiego Towarzystwa Geologicznego Kielce 6–8 września 1981*, 7–19.Wydawnictwa Geologiczne, Warszawa.

Kowalczewski, Z. 1990: Grubookruchowe skały kambru na środkowym południu Polski (litostratygrafia, tektonika, paleogeografia). *Prace Panstwowego Instytutu Geologicznego 131*. 82 pp.

Kowalczewski, Z. & Migaszewski, Z. 1993: Key problems of the tectonics and stratigraphy of the Palaeozoic rocks in the Holy Cross Mountains (Gory Swietokrzyskie), Poland. *In* Gee, D.G. & Beckholmen, M. (eds.): Europrobe Symposium Jablonna 1991. *Publications of the Institute of Geophysics, Polish Academy of Sciences A-20;255*, 99–104.

Kowalczewski, Z. Moczydłowska, M. & Kuleta, M. 1984: Uwagi o stratygrafii i tektonice skał kambryjskich nawierconych w podłożu Górnośląskiego Zagłębia Węglowego w otworach Goczałkowice IG 1, Sosnowiec IG 1 i *Potrójna* IG 1. *Kwartalnik Geologiczny 28*, 450–451.

Landing, E. 1992: Lower Cambrian of Southeastern Newfoundland: Epeirogeny and Lazarus faunas, lithofacies–biofacies linkages, and the myth of a global chronostratigraphy. *In* Lipps, J.H. & Signor, P.W. (eds.): *Origin and Early Evolution of the Metazoa*, 283–309. Plenum, New York, N.Y.

Le Hérissé, A. 1989: Acritarches et cystes d'algues Prasinophycees du Silurien de Gotland, Suede. *Palaeontographica Italica 76*, 57–302.

Lendzion, K. 1978. Charakterystyka paleontologiczna. *In* Areń, B. & Lendzion, K.: Charakterystyka stratygraficzno-litologiczna wendu i kambru dolnego. *Prace Instytutu Geologicznego 90*, 36–49.

Lendzion, K. 1983a: Rozwój kambryjskich osadów platformowych Polski. *Prace Instytutu Geologicznego 105*. 55 pp. (English abstract.)

Li Jun, 1987: Ordovician acritarchs from the Meitan Formation of Guizhou Province, south-west China. *Palaeontology 30*, 613–634.

Liñán, E., Álvaro, J., Gozalo, R., Gámez-Vintaned, J.A. & Palacios, T. 1995: El Cámbrico Medio de la Sierra de Córdoba (Ossa-Morena, S de España): trilobites y paleoicnologia. Implicaciones bioestratigraficas y paleoambientales. *Revista Española de Paleontología 10*, 219–238.

Liñán, E., Palacios, T., Villafaina, M., Gozalo, R. & Álvaro, J. 1993a: Middle Cambrian acritarchs in the levels with *Solenopleuropsis* and *Sao* (Trilobita) from Zafra (Badajoz Province, Spain). Biostratigraphical consequences. *Terra Abstracts, Abstract Supplement 6 to Terra Nova 5*, 3–4.

Liñán, E., Perejón, A. & Sdzuy, K. 1993b: The Lower–Middle Cambrian stages and stratotypes from the Iberian Peninsula: a revision. *Geological Magazine 130*, 817–833.

Link, P.K. 1995: Vendian and Cambrian paleogeography of the Western United States and adjacent Sonora, Mexico. *In* Rodrigez Alonso, M.D. & Gonzalo Corral, J.C. (eds.): *XIII Reunion de Geologia del Oeste Peninsular, Annual IGCP Project 319–320 Meeting, Comunicaciones*, 91–95. Signo, Salamanca.

Lister, T.R. 1970: A monograph of the acritarchs and Chitinozoa from the Wenlock and Ludlow Series and Millichope areas, Shropshire. Part I. *Palaeontographical Society Monographs 124*, 1–100.

Loeblich, A.R. Jr. 1970: Morphology, ultrastructure and distribution of Paleozoic acritarchs. *Proceedings of the North American Paleontological Convention September 1969. Part G*, 705–788.

Loeblich, A.R.Jr. & Drugg, W.S. 1968: New acritarchs from the Early Devonian (Late Gedinnian) Haragan Formation of Oklahoma, U.S.A. *Tulane Studies in Geology 6:4*, 129–137.

Loeblich, A.R.Jr. & Tappan, H. 1978: The Middle and Late Ordovician microphytoplankton from central North America. *Journal of Paleontology 52*, 1233–1287.

Loeblich, A.R.Jr. & Wicander, R. 1976: Organic-walled microplankton from the Lower Devonian Late Gedinnian Haragen and Bois d'Arc Formations of Oklahoma, U.S.A. Part I. *Palaeontographica B 159*, 1–39.

Lotze, F. 1945: Zur Gliederung der Varisziden in der Iberischen Meseta. *Geotektonische Forschung 6*, 78–92.

Lovelock, J.E. 1979: *Gaia: A New Look at Life on Earth*. 157 pp. Oxford University Press, Oxford.

Lovelock, J.E. 1992: The Gaia Hypothesis. *In* Margulis, L. & Olendzenski, L. (eds.): *Environmental Evolution. Effects of the Origin and Evolution of Life on Planet Earth*, 295–322. MIT Press, Cambridge, Mass.

Lovelock, J.E. & Margulis, L. 1974: Atmospheric homeostasis by and for the biosphere: The Gaia hypothesis. *Tellus 26*, 1–9.

Margulis, L. & Olendzenski, L. (eds.) 1992: *Environmental Evolution. Effects of the Origin and Evolution of Life on Planet Earth*. 405 pp. MIT Press, Cambridge, Mass.

Martin, F. 1977: Acritarches de Cambro-Ordovicien du Massif de Brabant, Belgique. *Institute royal des Sciences naturelles de Belgique, Sciences de la terre, Bulletin 51*, 1–33.

Martin, F. 1982: Some aspects of late Cambrian and early Ordovician acritarchs. *In* Basset, M.G. & Dean, W.T. (eds.): The Cambrian–Ordovician boundary: sections, fossil distributions, and correlations. *National Museum of Wales, Geological Series 3*, Cardiff, 29–40.

Martin, F. 1983: Chitinozoaires et Acritarches Ordoviciens de la plateforme du Saint-Laurent (Québec et sud-est de l'Ontario). *Geological Survey of Canada, Bulletin 310*, 1–59.

Martin, F. 1984: New Ordovician (Tremadoc) acritarch taxa from the middle member of the Survey Peak Formation at Wilcox Pass, southern Canadian Rocky Mountains. *Current Research, Part A, Geological Survey of Canada, Paper 84-1A,* 441–448.

Martin, F. 1992: Uppermost Cambrian and Lower Ordovician acritarchs and Lower Ordovician chitinozoans from Wilcox Pass, Alberta. *Geological Survey of Canada, Bulletin 420.* 57 pp.

Martin, F. 1993: Acritarchs: a review. *Biological Review 68,* 475–538.

Martin, F. & Dean, W.T. 1981: Middle and Upper Cambrian and Lower Ordovician acritarchs from Random Island, eastern Newfoundland. *Geological Survey of Canada, Bulletin 343,* 1–43.

Martin, F. & Dean, W.T. 1983: Late early Cambrian and early Middle Cambrian acritarchs from Manuels River, eastern Newfoundland. *Current Research, Part B, Geological Survey of Canada, Paper 83-1B,* 353–363.

Martin, F. & Dean, W.T. 1984: Middle Cambrian acritarchs from the Chamberlains Brook and Manuels River Formations at Random Island, eastern Newfoundland. *Current Research, Part A, Geological Survey of Canada, Paper 84-1A,* 429–440.

Martin, F. & Dean, W.T. 1988: Middle and Upper Cambrian acritarch and trilobite zonation at Manuels River and Random Island, eastern Newfoundland. *Geological Survey of Canada, Bulletin 381,* 91 pp.

Martin, F. & Yin Leiming 1988: Early Ordovician acritarchs from southern Jilin Province, north-east China. *Palaeontology 31,* 109–127.

McKerrow, W.S. & Cocks, L.R.M. 1995: The use of biogeography in the terrane assembly of the Variscan belt of Europe. *Studia Geophysica et Geodaetica 39,* 269–275.

Meilliez, F. & Vanguestaine, M. 1983: Acritarches de Cambrien Moyen et Superieur a Montcornet-en-Ardenne (France); premieres données et implications. *Comptes Rendus des Seances de l'Academie des Sciences, Serie 2: Sciences de la Terre 297,* 265-268.

Mendelson, C.V. 1993: Acritarchs and prasinophytes. *In* Lipps, J.H. (ed.): *Fossil prokaryotes and protists,* 77–104. Blackwell, Boston, Mass.

Mendelson, C.V. & Schopf, W. 1992: Proterozoic and selected Early Cambrian microfossils and microfossil-like objects. *In* Schopf, J.W. & Klein, C. (eds.): *The Proterozoic Biosphere. A Multidisciplinary Study,* 865–951. Cambridge University Press, Cambridge.

Mens, K., Bergström, J. & Lendzion, K. 1990: The Cambrian System on the East European Platform. *International Union of Geological Sciences, Publication 25.* 73 pp. Trondheim.

Mette, W. 1989: Acritarchs from Lower Paleozoic rocks of the western Sierra Morena, SW-Spain and biostratigraphic results. *Geologica et Palaeontologica 23,* 1–19.

Moczydłowska, M. 1988: New Lower Cambrian acritarchs from Poland. *Review of Palaeobotany and Palynology 54,* 1–10.

Moczydłowska, M. 1991: Acritarch biostratigraphy of the Lower Cambrian and the Precambrian–Cambrian boundary in southeastern Poland. *Fossils and Strata 29.* 127 pp.

Moczydłowska, M. 1993a. Is there Caledonian deformation in the TESZ (Trans-European Suture Zone) of Upper Silesia, southern Poland? *In* Gee, D.G. & Beckholmen, M. (eds.): Europrobe Symposium Jablonna 1991. *Publications of the Institute of Geophysics Polish Academy of Sciences A-20:255,* 119–122.

Moczydłowska, M. 1993b: Acritarch biostratigraphy of the Cambrian succession in Upper Silesia, southern Poland and its tectonic implications. *Geological Society of America, 1993 Annual Meeting Boston, Massachusetts, Abstracts with Programs,* 430.

Moczydłowska, M. 1995a: Neoproterozoic and Cambrian successions deposited on the East European Platform and Cadomian basement area in Poland. *Studia Geophysica et Geodaetica 39,* 276–285.

Moczydłowska, M. 1995b: Cambrian microplankton distribution in Iberia and Baltica, and possible palaeogeographic relationships. *In* Rodrigez Alonso, M.D. & Gonzalo Corral, J.C. (eds.): *XIII Reunion de Geologia del Oeste Peninsular, Annual IGCP Project 319–320 Meeting, Comunicaciones,* 117–120. Signo, S.L., Salamanca.

Moczydłowska, M. 1996a: Comments on 'Evidence for a Caledonian orogeny in Poland' by J.D. Johnston, J.A. Tait, G.J.H. Oliver and F.G. Murphy. *Transactions of the Royal Society of Edinburgh: Earth Sciences 86,* 227–230.

Moczydłowska, M. 1996b: Cambrian acritarch biochronology and the duration of acritarch zones. *Ninth International Palynological Congress Program and Abstracts, Huston, Texas, 1996,* 107.

Moczydłowska, M. 1997: Proterozoic and Cambrian successions in Upper Silesia: an Avalonian terrane in southern Poland. *Geological Magazine 134,* 679-689.

Moczydłowska, M. & Crimes, T.P. 1995: Late Cambrian acritarchs and their age constraints on an Ediacaran-type fauna from the Booley Bay Formation, Co. Wexford, Eire. *Geological Journal 30,* 111–128.

Moczydłowska, M. & Vidal, G. 1986: Lower Cambrian acritarch zonation in southern Scandinavia and southeastern Poland. *Geologiska Föreningens i Stockholm Förhandlingar 108,* 201–223.

Moczydłowska, M. & Vidal, G. 1988: How old is Tommotian? *Geology 16,* 166–168.

Moczydłowska, M. & Vidal, G. 1992: Phytoplankton from the Lower Cambrian Læså formation on Bornholm, Denmark: biostratigraphy and palaeoenvironmental constraints. *Geological Magazine 129,* 17–40.

Moczydłowska, M., Vidal, G. & Rudavskaya, V. 1993: Neoproterozoic (Vendian) phytoplankton from the Siberian Platform, Yakutia. *Palaeontology 36,* 495–521.

Molyneux, S. & Fensome, R.A. 1996: Nomenclatural note. A re-evaluation of the genus *Cristallinium* and its species. *In* Jansonius, J. & McGregor, D.C. (eds.): *Palynology: Principles and Applications 2,* 516–517. American Association of Stratigraphic Palynologists Foundation, Publishers Press, Salt Lake City, Utah.

Moore, T.B., Horodyski, R.J., Lipps, J.H. & Schopf, J.W. 1992: Distinctive problematical Proterozoic microfossils. *In* Schopf, J.W. & Klein, C. (eds.): *The Proterozoic biosphere. A multidisciplinary study,* 233–235. Cambridge University Press, Cambridge.

Naumova, S.N. 1960: Sporogo-pyltsevye kompleksy rifejskikh i nizhnekembrijskikh otlozhenij SSSR. [Spore-pollen assemblages of the Riphean and Lower Cambrian deposits in the USSR.] *In: Stratigrafiya pozdnego dokembriya i kembriya. Mezhdunarodnyj geologicheskij kongres 21 sesii. Doklady sovetskikh geologov,* 109–117. USSR Academy of Sciences, Moscow.

Oberc, J. 1977: Besteht ein kaledonisches tektogen in Südpolen? *Neues Jahrbuch für Geologie und Paläontologie, Mh 1,* 56–63.

Oberc, J. 1986: Historia ruchów paleozoicznych w południowo-zachodniej Polsce. *In: Historia ruchów tektonicznych na ziemiach Polskich,* 56–61. Wydawnictwo Uniwersytetu Wrocławskiego, Wrocław.

Ogurtsova, R.N. 1985: *Rastitelnye mikrofossilii opornogo razreza venda – nizhnego kembriya Malogo Karatau.* [*The plant microfossils of the Vendian – Lower Cambrian section in Maly Karatau.*] 135 pp. Ilim, Frunze.

Oliver, G.J.H, Corfu, F. & Krogh, T.E. 1993: U–Pb ages from SW Poland: evidence for a Caledonian suture zone between Baltica and Gondwana. *Journal of the Geological Society, London 150,* 355–369.

Orłowski, S. 1975. Lower Cambrian trilobites from Upper Silesia (Goczałkowice borehole). *Acta Geologica Polonica 25,* 377–383.

Paalits, I. 1992: Upper Cambrian acritarchs from boring core M-72 of North Estonia. *Proceedings of the Estonian Academy of Sciences, Geology 41,* 29–37.

Palacios, T. 1989: Microfosiles de pared organica del Proterozoico superior (Región Centralde la Península Ibérica). *Memorias del Museo Paleontologico de la Universidad de Zaragoza 3.* 91 pp.

Palacios, T. 1993: Acritarchs from the volcanosedimentary group Playon Beds, Lower–Upper Cambrian, Sierra Morena, southern Spain. *Terra Abstracts, Abstract Supplement 6 to Terra Nova 5,* 3.

Palacios, T. & Vidal, G. 1992: Lower Cambrian acritarchs from northern Spain: the Precambrian–Cambrian boundary and biostratigraphic implications. *Geological Magazine 129,* 421–436.

Palmer, A.R. & James, N.P. 1980: The Hawke Bay event: a circum-Iapetus regression near the Lower–Middle Cambrian boundary. *In* Wones, D.R (ed.): The Caledonides in the U.S.A., 15–18. *Virginia Polytechnic Institute State University Department of Geological Sciences, Memoir 2.*

Palmer, A.R. & Rowell, A.J. 1995: Early Cambrian trilobites from the Shackleton Limestone of the Central Transatlantic Mountains. *Journal of Paleontology 69, Supp. 6, The Paleontological Society Memoir 45*, 1–28.

Parnes, A. 1971: Late Lower Cambrian trilobites from the Timna Area and Har'Amram (Southern Negev, Israel). *Israel Journal of Earth Sciences 20*, 179–205.

Pharaoh, T., England, R. & Lee, M. 1995: The concealed Caledonide basement of Eastern England and the southern North Sea – a review. *Studia Geophysica et Geodaetica 39*, 330–346.

Pharaoh, T.C., Merriman, R.J., Webb, P.C. & Beckinsale, R.D. 1987: The concealed Caledonides of east England: preliminary results of a multidisciplinary study. *Proceedings Yorkshire Geological Society 46*, 355–369.

Pillola, G.L. 1990: Lithologie et trilobites du Cambrien inférieur de SW de la Sardaigne (Italie): implications paléobiogéographiques. *Comptes Rendus de l'Académie des Sciences, Paris, 310, Serie II*, 321–328.

Pittau, P. 1985: Tremadocian (Early Ordovician) acritarchs of the Arburese unit, southwest Sardinia (Italy). *Bolletino della Societá Paleontologica Italiana 23*, 161–204.

Playford, G. & Martin, F. 1984: Ordovician acritarchs from the Canning Basin, Western Australia. *Alcheringa 8*, 187–223.

Playford, G. & Wicander, R. 1988: Acritarch palynoflora of the Coolibah Formation (Lower Ordovician), Georgina Basin, Queensland. *In* Jell, P.A. & Playford, G. (eds.): *Palynological and Palaeobotanical Studies in Honour of Basil E. Balme. Association of Australasian Palaeontologists, Memoir 5*, 5–40.

Pocock, S.A.J. 1972: Palynology of the Jurassic sediments of western Canada. Part II. Marine Species. *Palaeontographica B, 137*, 85–153.

Pożaryski, W. 1990: Kaledonidy środkowej Europy orogenem przesuwczym złożonym z terranów. *Przegląd Geologiczny 1*, 1–9.

Repina, L.N., Lazarenko, N.P., Meshkova, N.P., Korshunov, V.L., Nikiforov, N.I. & Aksarina, N.A. 1974: *Biostratigrafiya i fauna nizhnego kembriya Kharaulakha (kh. Tuora–Sis).* [Biostratigraphy and fauna of the Lower Cambrian of Kharaulakh (ridge Tuora–Sis).] 299 pp. Nauka, Moscow.

Ribecai, C. & Vanguestaine, M. 1993: Latest Middle – Late Cambrian acritarchs from Belgium and northern France. *Special Papers in Palaeontology 48*, 45–55.

Riding, R. 1994: Evolution of algal and cyanobacterial calcification. *In* Bengtson, S. (ed.): *Early Life on Earth. Nobel Symposium No. 84*, 426–438. Columbia University Press, New York, N.Y.

Roberts, D. & Sturt, B.A. 1980: Caledonian deformation in Norway. *Journal of the Geological Society London 137*, 241-250.

Rovnina, L.V. 1981: Palynological method to determine the level of katagenesis of organic matter by using Jurassic deposits of western Siberia. *In* Brooks, J.G. (ed.): *Organic Maturation Studies and Fossil Fuel Exploration*, 427–432. Academic Press, London.

Rudavskaya, V.A. & Vasileva, N.J. 1984: Pervye nakhodki lukatiskikh akritarkh v nizhnem kembrii Chekyrovkogo rareza r. Leny. [The first finds of the Lükati acritarchs in the Lower Cambrian Chekurovsk section on the Lena River.] *Doklady Akademii Nauk SSSR 274*, 1454–1456. (In Russian.)

Rudavskaya, V.A. & Vasileva, N.J. 1985: Akritarkhi i skeletnaya problematika na granitsakh Venda, Tommotskogo i Atdabanskogo yarusov. [Acritarchs and the skeletal problematics at the boundaries of the Vendian, Tommotian and Atdabanian Stages.] *In* Kokoulin, M.L. & Rudavskaya, V.A. (eds.): *Startigraphy of the Late Precambrian and Early Palaeozoic of the Siberian Platform*, 51–57. VNIGRI, Leningrad. (In Russian.)

Sarjeant, W.S.A. 1968: Microplankton from the Upper Callovian and Lower Oxfordian of Normandy. *Revue de Micropaleontologie 10*, 221–242.

Sarjeant, W.S.A. 1985: A restudy of some dinoflagellate cyst holotypes in the University of Kiel Collections. VI. Late Cretaceous dinoflagellate cysts and other palynomorphs in the Otto Wetzel Collection. *Meyniana 37*, 129–185.

Sarjeant, W.S.A. & Stancliffe, R.P.W. 1994: The *Micrhystridium* and *Veryhachium* complexes (Acritarcha: Acanthomorphitae and Polygonomorphitae): a taxonomic reconsideration. *Micropalaeontology 40*, 1–77.

Schaarschmidt, F. 1963: Sporen und Hystrichosphaerideen aus dem Zechstein von Budingen in der Wetterau. *Palaeontographica B 113*, 38–91.

Scotese, C.R. & McKerrow, W.S. 1990: Revised world maps and introduction. *In* McKerrow, W.S. & Scotese, C.R (eds.): Palaeozoic Palaeogeography and Biogeography, 1–21. *The Geological Society London Memoir 12*.

Sdzuy, K. 1968: Biostratigrafia de la griotte cámbrica de Los Barrios de Luna (León) y de otras sucesiones comparables. *Trabajos de Geologia 2*, 45–57. Oviedo.

Sdzuy, K. 1971a: Acerca de la corelacion del Cambrico inferior en la Peninsula Iberica. *Actas I Congreso Hispano-Luso-Americano de Geologia Economica. Geologia 2*, 753–768. Madrid–Lisboa.

Sdzuy, K. 1971b: La subdivisiòn bioestratigráfica y la correlación del Cámbrico medio de España. *Actas I Congreso Hispano-Luso-Americano de Geologia Económica. Geologia 2*, 769–782. Madrid–Lisboa.

Sdzuy, K. 1972: Das Kambrium der Acadobaltischen Faunenprovinz. *Zentralblatt für Geologie und Paläontologie 2*, 1–91.

Seilacher, A. 1994: Early multicellular life: Late Proterozoic fossils and the Cambrian explosion. *In* Bengtson, S. (ed.): *Early Life on Earth. Nobel Symposium No. 84*, 389–400. Columbia University Press, New York, N.Y.

Slaviková, K. 1968: New finds of acritarchs in the Middle Cambrian of the Barrandian (Czechoslovakia). *Věstnik Ústředniho ústavu geologického 43*, 199–205.

Staplin, F.L. 1961: Reef-controlled distribution of Devonian microplankton in Alberta. *Palaeontology 4*, 392–424.

Staplin, F.L., Jansonius, J. & Pocock, S.A.J. 1965: Evaluation of some Acritarchous Hystrichosphere Genera. *Neues Jahrbuch für Geologie und Paläontologie. Abh. 123*, 167–201.

Stearn, W.T. 1983: *Botanical Latin. History, Grammar, Syntax, Terminology and Vocabulary.* 566 pp. David & Charles, Newton Abbot London North Pomfret.

Steiner, M. 1994: Die Neoproterozoischen Megaalgen Südchinas. *Berliner Geowissenschaftliche Abhandlungen E 15*, 1–146.

Stockmans, F. & Willière, Y. 1963: Les Hystrichosphères ou mieux les Acritarches du Silurien Belge. Sondage de la Brasserie Lust à Courtrai (Kortrijk). *Bulletin de la Société Belge de géologie, de paléontologie et d'hydrologie 71*, 450–481.

Stone, P. & Kimbell, G.S. 1995: Caledonian terrane relationships in Britain: an introduction. *Geological Magazine 132*, 461–464.

Sturt, B.A., Soper, N.J., Bruck, P.M. & Dunning, F.W. 1980: Caledonian Europe. *Episodes 1*, 13-16.

Sun Weiguo 1994: Early multicellular fossils. *In* Bengtson, S. (ed.): *Early Life on Earth. Nobel Symposium No. 84*, 358–369. Columbia University Press, New York.

Ślączka, A. 1975: Wyniki geologiczne otworu Potrójna IG 1. *Kwartalnik Geologiczny 19:2*, 487-488.

Ślączka, A. 1976: Nowe dane o budowie podłoża Karpat na południe od Wadowic. *Rocznik Polskiego Towarzystwa Geologicznego 46*, 337–350.

Ślączka, A. 1982: Profil utworow kambru w otworach położonych na południe wschód od Goczałkowic. *In* Różkowski, A. & Ślósarz, J. (eds.): *Przewodnik LIV Zjazdu Polskiego Towarzystwa Geologicznego, Sosnowiec 23–25 września 1982*, . Wydawnictwa Geologiczne, 201–205. Warszawa.

Ślączka, A. 1985a: Kambr. Eokambr. *In* Ślączka, A. (ed.): *Profile głębokich otworów wiertniczych Instytutu Geologicznego, Potrójna IG 1, Zeszyt 59*, 38 41. Wydawnictwa Geologiczne, Warszawa.

Ślączka, A. 1985b: Wyniki badan stratigraficznych i litologicznych. *In* Ślączka, A. (ed.): *Profile głębokich otworów wiertniczych Instytutu Geologicznego, Potrójna IG 1, Zeszyt 59*, 46–58. Wydawnictwa Geologiczne, Warszawa.

Ślączka, A. 1985c (ed.): *Profile głębokich otworów wiertniczych Instytutu Geologicznego, Potrójna IG 1, Zeszyt 59.* 166 pp. Wydawnictwa Geologiczne, Warszawa.

Tappan, H. 1980: *The Paleobiology of Plant Protists.* 1028 pp. Freeman, San Francisco, Cal.

Tappan, H. & Loeblich, Jr., A.R. 1971: Surface sculpture of the wall in Lower Paleozoic acritarchs. *Micropaleontology 17,* 385–410.

Timofeev, B.V. 1959: *Drevnejshaja flora Pribaltiki.* [*The oldest flora of the Baltic area.*] *Trudy VNIGRI 129.* 320 pp. Leningrad.

Timofeev, B.V. 1966: *Mikropaleofitologicheskoe issledovanie drevnykh svit.* [*Microphytological Research of Ancient Formations.*] 147 pp. Nauka, Moscow.

Tongiorgi, M. & Ribecai, C. 1990: Late Cambrian and Tremadocian phytoplankton (acritarchs) communities from Öland (Sweden). *Bolletino della Società Paleontologica Italiana 29,* 77–88.

Torsvik, T.H., Ryan, P.D., Trench, A. & Harper, D.A.T. 1991: Cambrian–Ordovician palaeogeography of Baltica. *Geology 19,* 7–10.

Tucker, R.D. & McKerrow, W.S. 1995: Early Paleozoic chronology: a review in light of new U–Pb zircon ages from Newfoundland and Britain. *Canadian Journal of Earth Sciences 32,* 368–379.

Turner, R.E. 1984: Acritarchs from the type area of the Ordovician Caradoc Series, Shropshire, England. *Palaeontographica B 190,* 87–157.

Tynni, R. 1982: On Paleozoic microfossils in clastic dykes on the Åland Islands and in the core samples of Lumparn. *Geological Survey of Finland, Bulletin 317,* 35–114.

Umnova, N.I. & Vanderflit, E.K. 1971: Kompleksy akritarkh kembrijskikh i nizhneordovikskikh otlozhenij zapada i severo-zapada Russkoj platformy. [Acritarch assemblages from the Cambrian and Lower Ordovician deposits of the western and northwestern Russian platform.] *In: Palinologicheskoe issledovaniya v Belorussii i drugikh rajonakh SSSR,* 68–73, Science and Engineering, Minsk.

Valensi, L. 1949: Sur quelques microorganismes planctoniques des silex du Jurassique moyen du poitou et de Normandie. *Société géologique de France, Bulletin 5e sér. 18,* 537–550.

Valentine, J.W. 1992: The macroevolution of phyla. *In* Lipps, J.H. & Signor, P.W. (eds.): *Origin and Early Evolution of the Metazoa,* 525–533. Plenum, New York, N.Y.

Valentine, J.1994: The Cambrian explosion. *In* Bengtson, S. (ed.): *Early Life on Earth. Nobel Symposium No. 84,* 401–411. Columbia University Press, New York, N.Y.

van Breemen, O., Bowes, D.R., Aftalion, M. & Żelaźniewicz, A. 1988: Devonian tectonothermal activity in the Sowie Gory gneissic block, Sudetes, southwestern Poland: evidence from Rb–Sr and U–Pb isotopic studies. *Annales Societatis Geologorum Poloniae 58,* 3–19.

Vanguestaine, M. 1974: Especes zonales d'acritarches de Cambro-Tremadocien de Belgique et de l'Ardenne Francaise. *Review of Palaeobotany and Palynology 18,* 63–82.

Vanguestaine, M. 1978: Criteres palynostratigraphiques conduisant a la reconnaissance d'un pli couche Revinien dans le sondage de Grand-Halleux. *Annales de la Société Géologique de Belgique 100,* 249–276.

Vanguestaine, M. 1991: Datation par acritarches des couches Cambro-Tremadociennes les plus profondes du sondage de Lessines (bord méridional du Massif de Brabant, Belgique). *Annales de la Société Géologique de Belgique 114,* 213–231.

Vanguestaine, M. 1992: Biostratigraphie par acritarches du Cambro-Ordovicien de Belgique et des regions limitophes: Synthese et perspectives d'avenir. *Annales de la Société Géologique de Belgique 115,* 1–18.

Vanguestaine, M. & Van Looy, J. 1983: Acritarches du Cambrian Moyen de la Vallee de Tacheddirt (Haut-Atlas, Maroc) dans le cadre d'une nouvelle zonation du Cambrien. *Annales de la Societe Geologique de Belgique 106,* 69–85.

Vasileva, N. 1985: Biostratigrafiya nizhnikh gorizontov kembriya severo-vostochnoj chasti sibirskoj platformy. [Biostratigraphy of the lower zones of the Cambrian in the northeastern part of the Siberian Platform.] *In* Kokoulin, M.L. & Rudavskaya, V.A. (eds.): *Stratigrafiya pozdnogo dokembriya i rannego paleozoya sibirskoj platformy. Stratigraphy of the Late Precambrian and Early Palaeozoic of the Siberian Platform,* 5–15. VNIGRI, Leningrad. (In Russian.)

Vavrdová, M. 1966: Paleozoic microplankton from Central Bohemia. *Časopis pro mineralogii a geologii 11,* 409–414.

Vavrdová, M. 1976: Excystment mechanism of Early Paleozoic acritarchs. *Časopis pro mineralogii a geologii 21,* 55–64.

Vavrdová, M. 1982: Recycled acritarchs in the uppermost Ordovician of Bohemia. *Časopis pro mineralogii a geologii 27,* 337–345.

Vavrdová, M. 1984: Some plant microfossils of possible terrestrial origin from the Ordovician of Central Bohemia. *Věstnik Ústředniho ústavu geologického 59,* 165–170.

Vidal, G. 1984: The oldest eucaryotic cells. *Scientific American 250,* 48–57.

Vidal, G. 1988: A palynological preparation method. *Palynology 12,* 215–220.

Vidal, G. 1994: Early ecosystems: Limitations imposed by the fossil record. *In* Bengtson, S. (ed.): *Early Life on Earth. Nobel Symposium No. 84,* 298–311. Columbia University Press, New York, N.Y.

Vidal, G. & Knoll, A.H. 1983: Proterozoic plankton. *Geological Society of America, Memoir 161,* 265–277.

Vidal, G. & Moczydłowska, M. 1992: Patterns of radiation in the phytoplankton across the Precambrian–Cambrian boundary. *Journal of the Geological Society, London 149,* 647–654.

Vidal, G. & Moczydłowska, M. 1995: Changes in Neoproterozoic–Cambrian phytoplankton biodiversity. *In* Rodrigez Alonso, M.D. & Gonzalo Corral, J.C. (eds.): *XIII Reunion de Geologia del Oeste Peninsular, Annual IGCP Project 319–320 Meeting, Comunicaciones,* 30–31. Signo, S.L., Salamanca.

Vidal, G. & Moczydłowska, M. 1996: Vendian – Lower Cambrian acritarch biostratigraphy of the central Caledonian fold belt in Scandinavia and the palaeogeography of the Iapetus–Tornquist seaway. *Norsk Geologisk Tidsskrift 76,* 147–168.

Vidal, G. & Moczydłowska, M. 1997: Biodiversity, speciation and extinction trends of Proterozoic and Cambrian phytoplankton. *Paleobiology 23,* 230–246.

Vidal, G., Moczydłowska, M. & Rudavskaya, V.A. 1995: Constraints on the early Cambrian radiation and correlation of the Tommotian and Nemakit–Daldynian regional stages of eastern Siberia. *Journal of the Geological Society, London, 152,* 499–510.

Vidal, G. & Peel, J.S. 1993: Acritarchs from the Lower Cambrian Buen Formation in North Greenland. *Grønlands Geologiske Undersøgelse, Bulletin 164.* 35 pp.

Volkova, N.A. 1968: Akritarkhi dokembrijskikh i nizhnekembrijskikh otlozhenij Estonii. [Acritarchs from the Precambrian and Cambrian deposits of Estonia.] *In* Volkova, N.A., Zhuravleva, Z.A., Zabrodin, V.E. & Klinger, B.Sh.: *Problematiki pogranichnykh sloev rifeja i kembriya Russkoj platformy, Urala i Kazakhstana,* 8–36. Nauka, Moscow.

Volkova, N.A. 1969a: Raspredelenie akritarch v razrezakh severo-vostochnoj Polshi. [Distribution of acritarchs in the north-eastern Poland.] *In* Rozanov, A.Yu. *et al.: Tommotskij yarus i problema nizhnej granitsy kembriya,* 74–76. Nauka, Moscow.

Volkova, N.A. 1969b: Akritarkhi severo-zapada Russkoj platformy. Acritarchs of the north-western Russian platform. *In* Rozanov, A.Yu. *et al.: Tommotskij jarus i problema nizhnej granitsy kembriya. Tommotian Stage and the Cambrian lower boundary problem* 224–236. Nauka, Moscow.

Volkova, N.A. 1974: Akritarkhi iz pogranichnykh sloev nizhnego-srednego kembriya zapadnoj Latvii. [Acritarchs from the transitional Lower–Middle Cambrian beds in the western Lotvia.] *In Biostratigrafiya i paleontologiya nizhnego kembriya Evropy i Severnoj Azii,* 194–262. Nauka, Moscow.

Volkova, N.A. 1980: Akritarkhi srednego i verkhnego kembriya Moskovskoj sineklizy. [Middle and Upper Cambrian acritarchs from the Moscow syneclise.] *Izvestiya Akademii Nauk SSSR Seriya geologicheskaya 12,* 49–57.

Volkova, N.A. 1983: Akritarkhi srednego i verkhnego kembriya severozapada Vostochno-Evropejskoj platformy. [Middle and Upper Cambrian acritarchs from the northwestern East European Platform.] *In:*

Stratigrafiya i korrelyatsiya osadkov metodami palinologii, 13–17. Akademia Nauk SSSR Sverdlovsk.

Volkova, N.A. 1990: *Akritarkhi srednego i verkhnego kembriya vostochno-evropejskoj platformy.* [*Middle and Upper Cambrian acritarchs form the East European Platform*]. 115 pp. Nauka, Moscow.

Volkova, N.A. 1993: Akritarkhi pogranichnykh otlozhenij kembriya i ordovika proglintovoj polosy Estonii (skvazhina M-56). [Acritarchs of the Cambrian–Ordovician transitional deposits of the cliff zone in Estonia (borehole M-56).] *Eesti Teaduste Akadeemia Toimetised, Geoloogia 41*, 15–22.

Volkova, N.A. & Golub, I.N. 1985: Novye akritarkhi verkhnego kembriya Leningradskoj oblasti (Ladozhskaya svita). [New acritarchs from the Upper Cambrian in the Leningrad area (Ladoga formation).] *Paleontologitcheskij Zhurnal 4*, 90–98.

Volkova, N.A., Kiryanov, V.V., Piscun, L.V., Pashkyavichene, L.T. & Jankauskas, T.V. 1979: Rastitelnye mikrofossilii. [Plant microfossils.] *In* Keller, B.M. & Rozanov, A.Yu. (eds.): *Paleontologiya verkhnedokembrijskikh i kembrijskikh otlozhenij Vostochno-Evropejskoj platformy*, 4–38. Nauka, Moscow.

Welsch, M. 1986: Die Acritarchen der hoheren Digermul-Gruppe, Mittelkambrium bis Tremadoc, Ost-Finnmark, Nord-Norwegen. *Palaeontographica B 201*, 1–109.

Westergård, A.H. 1950: Non-agnostidean trilobites of the Middle Cambrian of Sweden. II. *Sveriges Geologiska Undersökning C 511*. 57 pp.

Wetzel, O. 1933: Die in organischer Substanz erhaltenen Mikrofossilien des Baltischen Kreide-Feuersteins. *Palaeontographica A 78*, 1–110.

Wicander, E.R. 1974: Upper Devonian – Lower Mississippian acritarchs and prasinophycean algae from Ohio, U.S.A. *Palaeontographica B 148*, 9–43.

Wieser, T. 1985: Skały prekambryjskie. *In* Ślączka, A. (ed.): *Profile głębokich otworów wiertniczych Instytutu Geologicznego, Potrójna IG 1, Zeszyt 59*, 87–100. Wydawnictwa Geologiczne, Warszawa.

Wood, G.D. & Clendening, J.A. 1982: Acritarchs from the Lower Cambrian Murray Shale, Chilhowee Group, of Tennessee, U.S.A. *Palynology 6*, 255–265.

Woodcock , N.H. 1991: The Welsh, Anglian and Belgian Caledonides compared. *Annales Société Géologique Belgique 114*, 5–17.

Yin Lei-ming 1994: New forms of acritarchs from Early Ordovician sediments in Yichang, Hubei, China. *Acta Micropalaeontologica Sinica 11*, 41–53

Young, T., Martin, F., Dean, W.T. & Rushton, A.W.A. 1994: Cambrian stratigraphy of St Tudwal's Peninsula, Gwynedd, northwest Wales. *Geological Magazine 131*, 335–360.

Zamarreño, I. 1972: Las litofacies carbonatadas del Cámbrico de la Zona Cantabrica (NW. España) y su distribucion paleogeografica. *Trabajos de Geologia 5*. 118 pp. Oviedo.

Zang, W.-l. 1992: Sinian and Early Cambrian floras and biostratigraphy on the South China Platform. *Palaeontographica B 224*, 75–119.

Zang, W.L. & Walter, M.R. 1989: Latest Proterozoic plankton from the Amadeus Basin in central Australia. *Nature 337*, 642–645.

Zang, W. & Walter, M.R. 1992: Late Proterozoic and Cambrian microfossils and biostratigraphy, Amadeus Basin, central Australia. *Association of Australasian Palaeontologists Memoir 12*. 132 pp.

Zhuravlev, A. Yu. 1995. Preliminary suggestions on the global Early Cambrian zonation. *In* Geyer, G. & Landing, E. (eds.): *Morocco '95 The Lower–Middle Cambrian standard of Western Gondwana, Beringeria Special Issue 2*, 147–160. Würzburg.

Znosko, J. 1965: Pozycja tectoniczna śląsko-krakowskiego zagłębia węglowego. *Biuletyn Instytutu Geologicznego 188*, 73-98.

Znosko, J. 1984. Tectonics of southern part of Middle Poland (beyond Carpathians). *Zeitschrift der Deutschen Geologischen Gesellschaft 135*, 585–602.

Żaba, J. 1994: Mezoskopowe struktury kwiatowe w dolnopaleozoicznych utworach NE obrzeżenia GZW – rezultat transpresyjnego ścinania w strefie dyslokacyjnej Kraków–Myszków (Hamburg – Kraków). *Przegląd Geologiczny 42*, 643–648.

Żaba, J. 1995: Uskoki przesuwcze strefy krawędziowej bloków górnośląskiego i małopolskiego. *Przegląd Geologiczny 43*, 838–842.

Żelaźniewicz & Franke 1994: Discussion on 'U–Pb ages from SW Poland: evidence for a Caledonian suture zone between Baltica and Gondwana'. *Journal of the Geological Society, London, 151*, 1050–1052.

Öpik, A.A. 1979: Middle Cambrian agnostids: systematics and biostratigraphy. *Commonwealth of Austaralia Bureau of Mineral Resources, Geology and Geophysics Bulletin 172*. 188 pp.